ASTROBIOLOGY

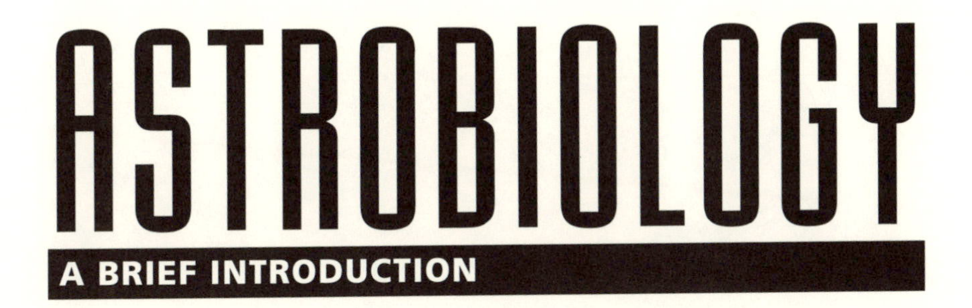

Kevin W. Plaxco and Michael Gross

The Johns Hopkins University Press Baltimore

Printed in the United States of America on acid-free paper
9 8 7 6 5 4 3 2 1

The Johns Hopkins University Press
2715 North Charles Street
Baltimore, Maryland 21218-4363
www.press.jhu.edu

Library of Congress Cataloging-in-Publication Data

Plaxco, Kevin W., 1965–
Astrobiology : a brief introduction / Kevin W. Plaxco and Michael Gross.
p. cm.
Includes bibliographical references and index.
ISBN 0-8018-8366-0 (hardcover : alk. paper) —
ISBN 0-8018-8367-9 (pbk. : alk. paper)
1. Life—Origin. 2. Exobiology. I. Gross, Michael, 1963– II. Title.
QH325.P57 2006
576.8′39—dc22 2005027406

A catalog record for this book is available from the British Library.

CONTENTS

PREFACE

Once upon a time, there was a scientific discipline called exobiology, the science of extraterrestrial life. Cynics said that it was the only research field with precisely no subject material to study. And thus, after a brief period of fame—thanks partly to the pioneering research and public outreach of Carl Sagan—exobiology fell out of fashion because it was too narrowly defined and focused on things that we did not, and indeed could not, know anything about.

Astrobiology, in contrast, removes the distinction between life on our planet and life elsewhere. It focuses on the more fundamental, and more tractable, question of the relationship between life and the Universe.* Not surprisingly, astrobiology tends to focus on life on Earth—Terrestrial life is, after all, the only example we have on hand—but it attempts to understand this single example in the broadest contexts of the Universe. Using our vast (but incomplete) knowledge of the Terrestrial biosphere, astrobiology addresses three broad questions about life in the Universe:

- What had to happen to allow the Universe to support life?
- How did the origins and evolution of life transpire on Earth, and how differently might they be transpiring elsewhere?
- Where else might life have arisen in our Universe, what might it be like, and how can we find it?†

Much of the worth of astrobiology lies in the fact that, by most standards, these are perhaps the most fundamental questions addressed by science today; they address the most profound issues regarding who we are and where we came from. Still more worth comes from the truly interdisciplinary approach that these questions demand; astrobiology

*Note that throughout this book we use *Universe, Solar System, Sun,* and *Moon* (and *Lunar*) for our particular universe, solar system, star, and moon; similarly, we use *Terrestrial* to refer to something of the Earth, *terrestrial* to refer to any rocky planet—though by *extraterrestrial* we mean specifically not of our Earth.

†And, in homage to the late Douglas Adams, author of *The Hitchhiker's Guide to the Galaxy,* "Will it buy me a drink?"

encompasses a variety of usually distinct scientific disciplines, ranging from cosmology, astrophysics, astronomy, geology, and chemistry to, of course, biology itself.

In this book we outline the current status of astrobiology with only the absolutely necessary amount of detail. We begin, in chapter 1, by defining the object of our study, life. While we can easily distinguish living from nonliving systems here on Earth, setting up a definition that encompasses all living systems in the Universe and unambiguously sets them apart from all nonliving systems requires careful consideration.

Next (chapter 2) we need to investigate how the stage was set for life to arise in our Universe. How did the Universe come into being, and which of the crucial factors in its origins and early history distinguish it from other possible universes that may not be able to harbor any life at all?

From the vastness of the Universe we zoom inward, in chapter 3, to explore the tiny blue dot that is our home planet and ask similar questions about it. Why did the third rock from the Sun turn into a living planet, while its neighbors did not? Which conditions are necessary for a planet—any planet, anywhere in the Universe—not only to become and remain habitable but also to give rise to life and thence a rich and diverse biosphere?

Zooming in by many more orders of magnitude, chapter 4 looks at the molecular world present at the surface of the young Earth and investigates which chemical conditions and which potential chemical pathways could have set the stage for life to originate here.

So the Universe, the Earth, and the molecules are all set and ready to go. But how did it actually happen? How did a habitable planet turn into an inhabited planet? The short answer is we don't know. Still, there are partial answers to some of our questions and constraints regarding others, giving us a chance to spin some speculative scenarios in chapter 5.

Following the history of life on our planet chronologically, in chapter 6 we move on from the first spark of self-replicating, evolving life to the first cells and organisms. Again, we know very little about what really happened, but our current knowledge allows us to put constraints on what may have happened here and risk an educated guess or two about how it might have to happen anywhere.

The veil begins to lift somewhat when we come to the last common ancestor of today's organisms, a bacterium that was already quite evolved and had DNA, RNA, and hundreds of distinct proteins. From that point onward molecular phylogeny and, increasingly, paleontology can help us to decipher the history of life on Earth as a proxy for life in our Universe, as we outline in chapter 7.

Is there life elsewhere in the Universe? Today, researchers are pondering this question with a much more concrete and, in some regards, more optimistic outlook than they did thirty years ago, when the *Viking* missions failed to detect life on Mars. This newfound optimism is based on the accumulating examples of organisms thriving in what we humans would consider extremely hostile conditions, such as at high pressures, in water well above its nominal boiling point, or in extremely salty, acidic, or alkaline brines. The study of these extremophiles, which we detail in chapter 8, has had serious repercussions: the discovery of life in many seemingly uninhabitable environments on Earth has radically expanded our perceptions of possible habitats on other planets. The search for life elsewhere in the Universe has thus become intimately connected to the search for new habitats on Earth, a consideration that constitutes an important element of the definition of astrobiology, setting it apart from the earlier, "extraterrestrial only" definition of exobiology.

Having explored the origins and the limits of life on Earth, in chapter 9 we expand our view beyond our home planet and ask: if life can thrive around hydrothermal vents in the deepest depths of our oceans, in the driest, coldest valleys of the Antarctic, and even deep within the Earth's crust, then which other places in the Solar System previously deemed inhospitable might actually be inhabited?

Ultimately, speculation on where, or even whether, extraterrestrial life exists frustrates the scientific mind, unless it can be followed up with actual exploration of the potential habitats concerned. Thus, in chapter 10, we conclude our brief overview of astrobiology with a survey of past, present, and future searches for our cosmic neighbors.

It is our hope that, by the end of this book, both cynics and enthusiast alike will be convinced that, unlike exobiology, astrobiology has a subject matter to study; its subject matter is nothing less than an understanding of our place in the Universe.

ACKNOWLEDGMENTS

We would like to thank the many friends and colleagues who were kind enough to edit our manuscript and just generally keep us on our toes—with such a highly interdisciplinary field on our hands, we couldn't possibly have done it on our own. Those deserving particular recognition include Stan Awramik, Jason Hollinger, George Karas, Roger Millikan, Stan Peale, John Perona, Susannah Porter, Véronique Receveur-Bréchot, Bill Sargent, Frank Spera, and Charlie Strauss, all of whom were kind enough to put time and effort into correcting our errors of fact and language. We would also like to thank Jayanth Banavar, Steve Benner, Dave Deamer, Frank Drake, Reza Ghadiri, Leslie Orgel, and Stan Peale for helpful discussions on aspects of their work, and Ken Oh and Brycelyn Boardman for preparing many of the figures.

Finally, Kevin Plaxco extends specific words of thanks to his Chem 147 classes, which were the direct inspiration for this work; his colleague Rob Geller, the designated "test audience" for the entire manuscript; his wife, Lisa Plaxco, who was supportive above and beyond the call of duty and who also edited every chapter save one (if you don't like chapter 6, we know who to blame); and his old friend Faiz Kayyem, for suggesting, nearly two decades ago, that "it'd be fun to write a book someday." He was right.

CHAPTER 1

What Is Life?

Erwin Schrödinger (1887–1961), reluctant cofounder of quantum mechanics, 1933 Nobel laureate in physics, and author of a famous thought experiment involving cruelty to felines, was used to speaking his mind. So much so that, after the Nazis came to power in 1933, he resigned from his chair at the University of Berlin, which he had taken over from Max Planck (1858–1947; 1918 Nobel laureate in physics) just six years earlier, and emigrated first to Oxford, then to his native Austria, from where he was exiled again after the *Anschluss.* In 1939, the government of neutral Ireland invited him to take up a chair of theoretical physics at the newly founded Dublin Institute for Advanced Studies. Although he was a political refugee, his landing was a soft one and he greatly benefited from his time at Dublin, where he remained for the next seventeen years.

One of the obligations of Schrödinger's new job was an annual public lecture. In 1943, he held a series of three lectures at Trinity College Dublin, where an audience of more than four hundred heard him discourse on the topic "What Is Life?" At a time when there was no such thing as biophysics, this venture of a theoretical physicist into the domain of biologists was unprecedented. Moreover, there was virtually nothing known in biology that would have satisfied the strict thinking of a physicist. So instead of giving answers, Schrödinger formulated some fundamental questions of biology, as seen by a physicist.

Schrödinger mainly covered two fundamental aspects of life, namely heredity and thermodynamics. He framed these in the basic questions of how life creates "order from order" and how it creates "order from disorder." In his analysis of genetics (order from order), he estimated the number of atoms contained in a gene (then a highly abstract concept). He proposed that the genetic information might be encoded in something resembling an aperiodic crystal—that is, a combination of a regular structure with information-bearing variations—an idea that, in retrospect, seems startlingly prescient. In the second half of his discourse, Schrödinger clarified that organisms can create ordered arrangements of molecules (and cells and tissues, in the case of higher organisms) within themselves, by creating even greater disorder in the environment. Thus was the evolution of highly complex organisms from a chaotic pool of simple, lifeless chemicals kept in line with the second law of thermodynamics.

Ultimately Schrödinger's lectures were published as a small book—which is still in print today—that was hugely influential. For the first time a prominent scientist had raised the question of how the *physics* of our Universe fundamentally constrains its *biology.* Still, in 1943 the question "what is life?" was wide open and posed major challenges not just to biology but across all of science. In the six decades that have passed since then, many aspects of this question have been resolved, such that today, in this opening chapter, we can take a stab not only at defining what life is but also at listing some of its most fundamental requirements.

Life

So, with the knowledge we have at the beginning of the twenty-first century, what, precisely, is life? Most answers to this question are reminiscent of the claim of U.S. Supreme Court Justice Potter Stewart (1915–85) that, while he could not precisely and unambiguously define pornography, "I know it when I see it." But that kind of empirical approach, of course, does not get us very far if we are going to embark on a deep and rigorous evaluation of the origins of life and its relationship to the Universe at large.

Moreover, if we are interested in what *might* have happened—the range of possibilities that could have unfolded on Earth, or might be happening elsewhere in the Universe—we have to attempt to define the boundary conditions of life. That is, we need to attempt to understand the range of conditions and events that had to conspire to make life possible.* In this, as perhaps in the definition of life itself, we must necessarily be somewhat parochial; our understanding of the conditions under which life can arise and evolve is almost certainly going to be flavored by deeply held preconceptions based on our understanding of life on Earth. But as long as we are aware of this underlying bias, we can at least tackle each of the seemingly necessary conditions in as unbiased and logical a fashion as is (even the word itself is telling) *humanly* possible.

Definition of Life

Life scientists should know what the first word in their job title means, but practitioners of various disciplines ranging from the origins of life

*Keep in mind, too, that evolution is quite good at generating organisms that are fit to survive under very harsh conditions. And thus the set of conditions under which life can thrive is probably much broader than the set of conditions under which it can arise from inanimate matter in the first place.

through to modern astrobiology have consistently failed to come up with an all-inclusive definition of life. Nevertheless, we shall conjure up a working definition of life that, if not perfect, will suffice as a basis for our discussion.

The most striking property that distinguishes living systems from the inanimate world is their ability to copy themselves, a process scientifically described by the term *self-replicating*. Among *Homo sapiens* the process is more colloquially captured in the phrase "get married, settle down, and have kids." The fact that living things copy themselves is so central to all of biology that some wag once pointed out that "life is just a DNA molecule's way of reproducing itself." All of biology, from bacterial mats through to warring nations, can be described as tools for or consequences of the replication of genes.

Another key limit to our discussion is to define life as a *chemical system* (as opposed to a mechanical or electronic system). Over decades, writers of science fiction and of putative nonfiction extrapolating current (nano)technological trends into the future have suggested that self-replicating, microscopically small robots will soon be cleaning out our arteries, degrading toxic waste, and generally making themselves useful. Irrespective of the accuracy of these predictions, it seems likely that physical laws of the Universe allow the creation of mechanical beings that can construct copies of themselves and thus meet our first criterion for life. Similarly, there are viable organisms in cyberspace, known as worms. While they require a computer, an internet connection, and, typically, some poorly written software in order to reproduce, one might argue that these items constitute their ecosystem; we are just as critically reliant on our ecosystem for reproduction. When working out a universal definition of life, it's not so easy to dismiss these potentially living things out of hand. Especially when, as technology progresses, the boundaries between biological, mechanical, and electronic systems will probably slowly erode, as brains will be interfaced with computers and microrobots will resemble insects.

Considering the origins and distribution of life in the Universe, however, it is difficult to imagine that mechanoid life (much less life dependent on the existence of an internet) would have arisen spontaneously. The problem is that mechanical things, by definition, use parts that are larger than molecules (if a system consists of molecular-scale parts, then it is by definition a *chemical* system). Before the creation of the first organisms, these parts would have to be moved around by the random fluctuations of solvent molecules moving to and fro, which is called Brownian motion. Brownian motion isn't all that fast: if you gently open a bottle of perfume, how long does it take Brownian motion to waft the molecules of fragrance across the room? And since thermal motions vary with the square root of the mass of the diffusing object, it would take far longer than the age of the Universe for a bucket of watch parts to spontaneously assemble into a watch, much less into a machine capable of copying itself. Thus, while mechanically based life might arise by the intelligent design of chemical life forms, it seems unlikely that it can arise spontaneously. In a nutshell, it seems fair not to worry about whether the self-replicating robots of the planet Lexus Nine are alive.

A final, but critical, element in our definition of life emerges from the observation that not all self-replicating chemical systems are alive. For example, crystals are, in a sense, self-replicating. This is particularly true in a supersaturated solution. Under such conditions, if one were to smash a growing crystal into smaller pieces, each of the pieces would in turn grow into a new and larger crystal. Crystals even breed true. For example, whereas individual molecules of sodium chlorate are not "handed" (they are superimposable on their mirror images—more on this in chapter 5), crystals of this substance do have a handedness (chirality). When you allow a sodium chlorate solution to crystallize, half of the crystals will be the mirror image of the other half: half left handed, the other half right. However, if you take a supersaturated solution of sodium chlorate and stir vigorously while it begins to crystallize, all of the crystals that form will be either right or left handed. Why is this? It is because the vigorous stirring shatters the first crystal that forms, and the minicrystals thus formed nucleate the growth of all the crystals that follow, causing them to adopt the same handedness. Crystallization is self-replication. Equally clearly, though, while "a diamond is forever," it is not and, with rare exceptions,* never was alive. So what do we need to add to our definition of life in order to discriminate between inanimate crystals and truly animate chemical systems? In a word, evolution.

Living beings produce offspring in their own image by the replication of their genetic material. But the replication is not perfect: random genetic mutations produce inheritable differences that may improve or impede an offspring's viability. These give natural selection a chance to shape the fate of future generations and, indeed, the evolution of a species. Evolution, with the inherent adaptation under selective pressures, is a fundamental property of life and clearly distinguishes it from inanimate, if sometimes self-replicating, materials. A crystal makes per-

*"Rare exceptions," you dare ask? In the 1950s, shortly after General Electric developed a commercially viable method for the production of industrial diamonds, the technicians there synthesized some out of peanut butter. Upping the ante, the company LifeGems now provides the service of converting some of the carbon in your deceased "loved ones" into permanent and rather sparkly keepsakes.

fect copies of itself. The first crystal of quartz that condensed out of the solar nebula 4.57 billion years ago is identical to the quartz that crystallized last week in the Corning Glassware plant in upstate New York. Crystals and crystallization are changeless, incapable of evolving into new, more complex, and better forms. And thus they are not, and never have been, alive.

Limitations of Our Definition

So, there we have our definition. Life is a self-replicating chemical system capable of evolving such that its offspring might be better suited for survival. As definitions go, this one is nice, clean and concise. Too bad it is fatally flawed. Or at least seriously limited. While a chemical system that is capable of reproduction and evolution is clearly alive, the reverse is not necessarily true; many things that fail to meet these criteria are obviously *also* alive. Those of us whose child-bearing years are past, for example, might take umbrage at the suggestion that they are not alive simply because they are no longer reproducing. But while it would be nice to have a definition of life that could be applied as a litmus test to every single specimen and include even post-reproductive academics, it is not necessary for our discussion. In many regards, evolution acts on the level of populations and species. For a species to thrive it must have individuals capable of reproducing, but there may very well be members of the species that serve its survival without reproducing at all, as is true for most members of ant or bee colonies. Thus, by defining a living organism as a self-replicating, evolving chemical system, we cover all species known to date (if not all individuals). Moreover, since replicating organisms must have preceded any given nonreplicating organism, this definition is sufficient for our needs because it does not artificially constrain our discussion of the origins and evolution of life.

Requirements for Life

What, then, are the fundamental conditions that life requires? Given that our knowledge of this subject is necessarily parochial, we should cast our net wide, making an effort not to mistake Terrestrial constraints for universal ones. Still, there are a number of criteria that seem to be absolutely critical elements for the formation of life.

Life requires chemistry. This means that life requires atoms more complex than hydrogen, whose solo chemistry is limited to the formation of H_2, and helium, which is one of the few chemical elements that lack any chemistry whatsoever. Even taking into account our potentially parochial, Terrestrial biases, it seems fairly certain that a self-replicating chemical system cannot be built using just the reaction $H + H \rightarrow H_2$.

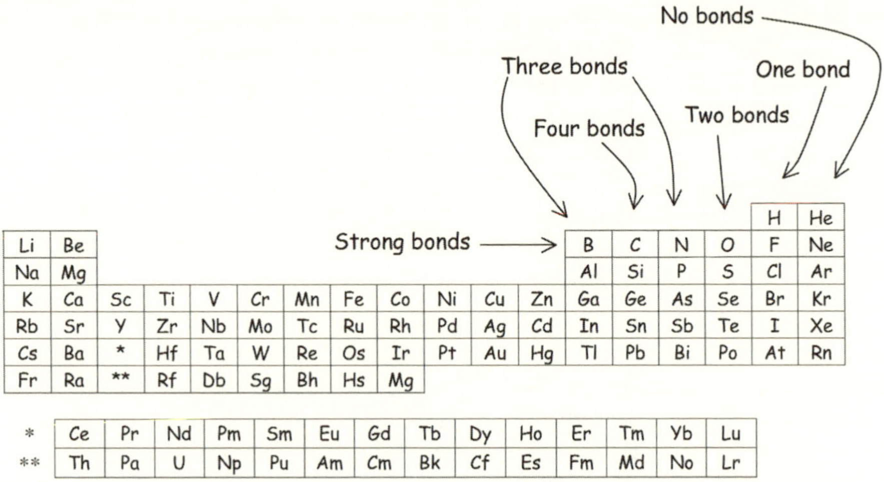

No bonds
Three bonds
One bond
Four bonds
Two bonds
Strong bonds

																H	He
Li	Be											B	C	N	O	F	Ne
Na	Mg											Al	Si	P	S	Cl	Ar
K	Ca	Sc	Ti	V	Cr	Mn	Fe	Co	Ni	Cu	Zn	Ga	Ge	As	Se	Br	Kr
Rb	Sr	Y	Zr	Nb	Mo	Tc	Ru	Rh	Pd	Ag	Cd	In	Sn	Sb	Te	I	Xe
Cs	Ba	*	Hf	Ta	W	Re	Os	Ir	Pt	Au	Hg	Tl	Pb	Bi	Po	At	Rn
Fr	Ra	**	Rf	Db	Sg	Bh	Hs	Mg									

*	Ce	Pr	Nd	Pm	Sm	Eu	Gd	Tb	Dy	Ho	Er	Tm	Yb	Lu
**	Th	Pa	U	Np	Pu	Am	Cm	Bk	Cf	Es	Fm	Md	No	Lr

FIG. 1.1. Perusal of the periodic chart draws us to the conclusion that relatively few atoms are likely to support the complex chemistry required to form life, here or anywhere. Only the second-row elements produce strong bonds with themselves (e.g., carbon-carbon or carbon-oxygen bonds), and thus only these atoms can be arranged in the type of large, complex molecules almost certainly required to generate a self-replicating chemical system. In particular, carbon is unique among the ninety or so naturally occurring elements in that it can form up to four strong covalent bonds with itself. Given the advantage this provides in terms of complex chemistry, creating life out of elements other than carbon would seem a significantly greater hurdle.

Thus the formation and evolution of chemical life will require atoms more complex than the two lightest atoms. And, as we will see, while hydrogen and helium were formed in great abundance in the first minutes of the Universe, the formation of heavier atoms was a far more delicate matter.

What atoms are required for life? Here we are perhaps on shakier ground, but not much shakier. Even a quick glance at the periodic table (fig. 1.1) shows that there are only a finite number of atoms out of which life could possibly be built. Do any of them have properties that uniquely suit them for the formation of life? The answer may well be yes.

It seems a fair assumption that a chemical system capable of copying itself will require at least a modest degree of complexity, and building complex molecules requires that we bond many atoms together. Clearly this cannot be done for the noble gases helium (He), neon (Ne), and argon (Ar), as these atoms do not participate in any chemistry. Nor can we build a complex chemistry based on atoms, such as chlorine, that make only one bond; at best they can form diatomics such as the aforementioned H_2. Thus in our search for atoms that could serve as the framework chemistry of life we can discount the first, second-to-last, and last columns of the periodic table, which are filled with such "uninteresting" atoms.

Similarly, to serve as the foundation of complex molecules, an atom must form very strong bonds to other atoms, and probably to itself (more precisely, to another atom of the same type). What do we mean by strong? We mean bonds that are hundreds of times stronger than the energy contained in a typical molecular collision, lest these same collisions tear the molecules apart. As we go down the periodic table, the outer electrons in each succeeding row of atoms—the electrons that participate in bonds—are more and more weakly bound to the nucleus. This occurs because each succeeding row in the table represents another filled shell of electrons, and with each row the outer electrons are more and more shielded from (i.e., less and less attracted to) the positively charged nucleus. Because of this, the bonding strength of the second, third, and fourth rows of the periodic table becomes progressively weaker. This is a serious issue. Whereas carbon, boron, and the like make for long, extremely strong chains of molecules (e.g., the long polymer chains that plastics are made of), no one has ever made a chain of silicon that was more than two atoms long; the Si–Si bond is simply too weak. Only the second-row elements are capable of forming strong covalent bonds to one another and to elements in the other rows. Thus we are probably on fairly strong ground in discounting all but this second row of elements in our quest to find the minimum set of materials necessary to form life.

There is one last criterion that might segregate reasonable life-forming elements from those that are much less likely to participate in the process: abundance. Even a casual glance around the Earth suggests that some elements are much more precious than others; gold is expensive because it is rare, whereas oxygen costs just a few cents per kilogram in industrial quantities. We go into this in detail in chapter 3, so let it suffice to say here that among the eight elements in the second row of the periodic table, lithium, beryllium, boron, and fluorine are relatively rare in the Universe (fig. 1.2). Thus a theory about alternative life forms that relies critically on these elements is significantly more suspicious than one that does not.

Life based on molecules almost certainly requires a *solvent* in which to move them around. Because mass transport through solids is at best extremely slow, solid-phase chemistry is far too limited to support the

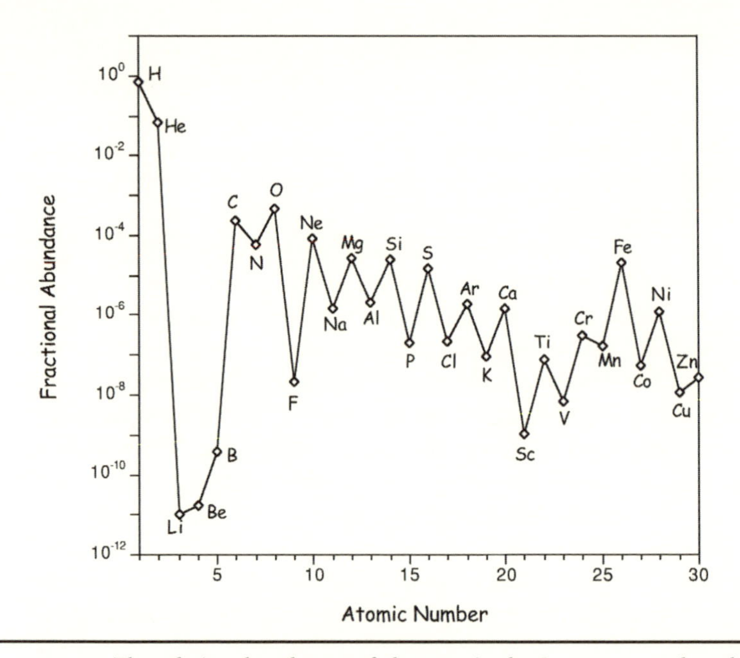

FIG. 1.2. The relative abundances of elements in the Sun map out the relative abundances available in the presolar nebula and thus, in turn, available to serve as the basic building blocks of life. The relative abundances of elements such as carbon and oxygen render them inherently more likely to serve as the basis of life than the rarer elements. Of note, the Sun is significantly enriched in elements heavier than helium relative to other stars its size. Nevertheless, this qualitative pattern is common among stars of the Sun's generation throughout the Universe (more on this in the next chapter).

complex networks of chemistry required for a self-replicating organism. This observation once again highlights the unique ability of the second-row elements to support life; Terrestrial animals eat water-soluble carbon compounds, such as the sugar and other carbohydrates in your morning doughnut, oxidize them with water-soluble oxygen, and excrete equally water-soluble carbon dioxide; silicon-based life forms, in contrast, would have a much more difficult time exhaling silicon dioxide, which tends to appear in solid forms, such as sand.

The human body contains around 70% water, highlighting the fact that, for Terrestrial organisms, the solvent in question is water. But is water the only plausible solvent for life? Once again a quick glance at the periodic table suggests that out of the (very finite) list of potential "biotic solvents" water may well be the only reasonable option. Water has so many properties that render it ideally suited as a biological solvent that its ability to form the basis of biochemistry may well be unique.

Some of the "ideal" properties of water are well known, and others, while less so, are no less critical for life on Earth. An example is taught to almost every elementary school student: water is one of the very few substances that expand when they freeze. Because of this, ice floats. If, instead, the ocean were filled with liquid ammonia, its winter pack "ice" would sink, where it would be insulated from the summer's warmth and prevented from seasonally melting. With each passing year, more and more of the ocean's volume would be locked up in the solid until, in a timeframe quite rapid by geological standards, the planet would freeze over, with only a thin seasonal layer of liquid on the surface. Could such a frozen ocean support the origins of life? Perhaps. But a permanently liquid ocean, with its ability to transport nutrients and modulate temperature, seems more likely to do the trick.

Water also has an extraordinary ability to absorb heat without much of a rise in its temperature, which is why we use it as a carrier for heat in central heating systems and hot water bottles and, conversely, also as a coolant. In more precise terms, the heat capacity of liquid water is 1 cal/g °C (by definition, it takes precisely 1 calorie—that is, 4.184 joules —to heat 1 gram of water from 14.5°C to 15.5°C). This value is about three times higher than that of typical rock or metal. This high heat capacity helps to moderate the Earth's climate, a seemingly critical event in the origins and evolution of life that we will cover in more detail in chapter 3. On a related note, thanks to its unique ability to form extended hydrogen-bonding networks, water remains liquid over a surprisingly broad, hundred-degree Celsius temperature range, thus helping to ensure that, even if the climate does fluctuate radically, liquid solvent will be available for life.

In addition to these important physical properties, the chemical properties of water seem to render it ideally suited as the basis for life. For example, the dielectric constant of water is around 80, which is significantly higher than that of any other cosmologically abundant liquid. This means that two oppositely charged ions in water are attracted with one-eightieth the force they would feel in a vacuum. Because of this, water can shield charged ions from one another, allowing them to be readily taken into solution, where they can perform chemistry. Water also has the highest molar density of any molecular liquid; fully 55.5 moles of water (3.3×10^{25} molecules) are crammed in each and every liter of the stuff. No other liquid packs anywhere near this many molecules in a given volume. Because of this extraordinary molar density, the entropy cost of "organizing" water ("the solvent") around any molecules dissolved in it ("the solute") is quite high (many water molecules need to be moved out of the way to make room for each cubic nanometer of solute), and thus water tends to force many types of dissolved molecules to organize themselves in order to minimize this entropic cost. This organizing effect, which is called the "hydrophobic effect,"* plays a critical role in organizing biomolecules on Earth. Lastly, water is cosmologically abundant, as its components, hydrogen and oxygen, are the first and third most abundant atoms in the Universe (fig. 1.2).

Of course, the fact that water is well suited for life on Earth doesn't automatically rule out that life elsewhere might be based on some other solvent. Or does it? It would be hard to find an alternative, as no other liquid has even a fraction of the favorable attributes of water. Hydrogen fluoride perhaps comes closest. Compared with water, it has a slightly higher dielectric constant (84, to water's 80), and thus it is at least as good a solvent for ionic materials. It also has a slightly wider, 102 degrees Celsius, liquid range (at atmospheric pressure it freezes at −83°C and boils at 19°C) and a comparable molar density (48.0 versus 55.5 mol/L). But as fluorine is cosmologically rare—it is about 1/100,000 as abundant as oxygen—it seems very unlikely that there are little purple fish happily swimming in seas of liquid HF on the planet Zap Seven.

Thus we are probably safe ruling out hydrogen fluoride. And none of the other molecular liquids formed by the cosmologically abundant elements (such as ammonia, hydrogen sulfide, or methane) comes anywhere near as close to the ideal properties of water as does HF; their liquid ranges are extremely small, their ability to solvate ionic materials is poor to effectively nonexistent, and their ability to regulate climate is extremely limited (table 1.1). From such considerations emerges the near certainty not only that life has an absolute requirement for a liquid solvent, but that water is by far the most "qualified" solvent to fulfill that role. This is not to say that life cannot have arisen based on other solvents; simply that the origins of life face a much more significant hurdle in the absence of this remarkable and abundant liquid.

Life also requires a solid or liquid *substrate.* The reason that life probably cannot exist in the gas phase is that molecules of sufficient complexity to form life are inevitably too dense to stay suspended. This is, of course, a much bigger constraint on the origins of life than on its ability to thrive after it has arisen; if the surface of the Earth were slowly to become uninhabitable, it is a pretty good bet that at least some bacteria would adapt to full-time living in cloud droplets. Indeed, as we discuss

*It's why, for example, "oil and water don't mix."

TABLE 1.1
Physical properties of potential biological solvents

Solvent	Formula	Liquid range (°C, at 1 atmosphere)	Molar density (mol/L)	Heat capacity (cal/g °C)	Dielectric constant
Water	H_2O	0 to +100	55.5	1.0	80
Hydrogen fluoride	HF	−83 to +19	48.0	0.8	84
Ammonia	NH_3	−78 to −34	40.0	1.1	25
Hydrogen sulfide	H_2S	−85 to −6	26.8	0.5	9
Methane	CH_4	−182 to −161	26.4	0.7	2
Hydrogen	H_2	−259 to −253	35.0	0.002	1

in a later chapter, some may already have done so. But the limited mass transport and limited size of condensed bodies that can occur in the gas phase make this realm an exceedingly unlikely one for the *origins* of life. This effectively rules out life on Jupiter (if life couldn't get started in the first place, it was unlikely to have evolved into giant, hydrogen-filled Hindenburgoids), much less, for example, in interstellar space.*

And let us not forget that life requires *energy.* This is obvious for the chemist, as living organisms create an implausible amount of order out of disorder, such as when the randomly distributed molecules of carbon dioxide and fertilizers end up in the highly nonrandom structure of a plant. According to the second law of thermodynamics, they can achieve this only if at least as much entropy (roughly speaking, a measure of molecular disorder) is created elsewhere. By using energy from an external source, life can swap energy for entropy: the living organisms get the calories and the order, while the rest of the Universe pays the price.

Thus, life requires an external disequilibrium (an "ordered" state) whose tendency to drive chemical reactions toward a more equilibrated ("disordered") state it can exploit for its own purpose of organizing its molecules into some pattern capable of reproduction. Among the most abundant sources of disequilibrium in the Universe are temperature differences: the fact that stars are much hotter than the Universe at large. Because of this, the copious number of high-energy photons emitted by a star can be absorbed by the surface of a (much cooler) planet. Here on

*This didn't stop the late cosmologist Fred Hoyle from writing a wonderful bit of science fiction, *The Black Cloud,* about an intelligent interstellar cloud that pays the Solar System a visit and accidentally wipes out most of humanity. It simply hadn't occurred to the cloud that life was possible on the cold, densely packed surface of a planet.

Earth, plants take advantage of this disequilibrium and use it to feed the striking disequilibrium that is our biosphere. For example, the presence of combustible wood in an atmosphere containing oxygen is a clear deviation from chemical equilibrium with respect to a mixture of water and carbon dioxide, as forest fires remind us. We animals, in turn, take advantage of the latter disequilibrium when we oxidize the carbohydrates in our morning doughnut to generate the energy we use to run our metabolic processes.

Substrates and solvents and thermodynamics aside, life also presumably requires *time.* And the narrower the range of conditions under which life can arise in the first place, the more unlikely will be the occurrence of a sufficiently stable environment that will stay within the range for sufficiently long. The Universe is a dangerous place. The luminosity of a star changes, and with it the temperature of any planets warmed by its light. Planets are sometimes struck by asteroids so large that the energy imparted by the impact can boil oceans and sterilize worlds. Atmospheres escape into space. Rotational axes tilt, plunging planets into million-year winters. Supernovae explode with the power of a billion suns, sterilizing any planet within a few hundred light-years. Considering these risks, it is clear that not all of the environments in the Universe that are capable of supporting the formation of life will remain stable long enough for life to arise at all, much less gain a secure footing.

Conclusions

So the recipe for life to arise somewhere in the Universe seems relatively straightforward. All we need is some water, carbon on a solid (or liquid) planetary surface, an energy source, some time, and we're off. But is it that easy? What is required to produce a water-and-carbon-bearing planetary environment that provides energy sources and yet is stable over eons? And how often are these conditions met? And if we find these conditions, how likely is it that life will arise? The following chapters explore each of these critical questions in turn.

More than fifty years after Schrödinger's lectures, the most fundamental aspects of the questions he asked have been answered, even if some details are left to fill in. In the summer of 1993, a dozen prominent scientists, including Nobel laureates Christian de Duve and Manfred Eigen, science popularizers Jared Diamond and Stephen Jay Gould (1941–2002), and evolutionary pioneers John Maynard Smith and Leslie Orgel, assembled at Trinity College Dublin to commemorate the lectures and to deliver new lectures. Their ambitious goal was to set a research agenda for the next fifty years of life science research, as Schrödinger had done. Because the investigation of present life on our planet has become relatively straightforward, many of the lectures focused on the mysteries of the origins and early evolution of life on Earth, which will also loom large in the chapters around the middle of this book.

Further Reading

Expectation of and constraints on life in the Universe. Schulze-Makuch, Dirk, and Irwin, Louis. *Life in the Universe.* Berlin: Springer-Verlag, 2004.

The history of origins-of-life research. Fry, I. *The Emergence of Life on Earth.* New Brunswick, NJ: Rutgers University Press, 2000.

CHAPTER 2

Origins of a Habitable Universe

The first communications satellite to orbit our planet was just a mirror. Launched in 1960, *Echo* consisted of a metalized balloon, about 70 meters in diameter, that simply reflected radio waves from a transmitter on one side of the Atlantic to receivers on the other. After a few years, the passive *Echo* was replaced by the first "active" communications satellites (satellites that detect and electronically amplify signals before sending them on to the recipient), obviating the need for the highly sensitive antennas and receivers built for the earlier, passive system.

In 1965 Arno Penzias and Robert Wilson, radio physicists working for AT&T's Bell Labs in New Jersey, realized that a semi-retired radio receiver built for the *Echo* program and conveniently located at nearby Crawford Hill might be of use for the astronomical detection of radio waves emanating from our galaxy. To accurately characterize these presumably very faint signals, their first task was to eliminate the various sources of electronic noise that plague any instrument. They pointed the 6.1-meter diameter, horn-shaped antenna at what were assumed to be "empty" (radio-silent) parts of the sky to measure this background noise. As expected, even in these empty regions of the sky they detected a faint radio-frequency "hiss" that they assumed arose from instrument artifacts. Penzias and Wilson systematically set about "fixing" the antenna and its associated amplifiers, eliminating one by one the sources of electronic static. But the noise persisted. Eventually, after having tested every conceivable electronic source, the physicists came to suspect that a pair of pigeons that had roosted in the horn might be the source of the offending signals, but neither chasing the pigeons away nor cleaning up years' worth of pigeon droppings (the life of a physicist is not always as glamorous as it appears in the movies) cleared up the annoying hiss. They were flummoxed. Then they heard rumor of a not-yet-published paper by three physicists at nearby Princeton University, Jim Peebles, David Wilkinson (1935–2002), and Bob Dicke (1917–97). The paper outlined an important prediction of a theory of the origins of the Universe. This theory, they wrote, predicted that the entire Universe would be filled with a pervasive hum of radio-frequency radiation (now termed the *cosmic microwave background*) at precisely the frequency and intensity observed in the horn antenna.

Unbeknownst to Peebles, Wilkinson, and Dicke, a very similar theory had been described as far back as 1948 by the Hungarian-born American physicist George Gamow (1904–68) and his student Ralph Alpher. Together they postulated that the Universe might have formed from an initially super-dense, super-hot state from which, today, billions of years later, it continues to expand and cool. They based this theory, in part, on observations made in the late 1920s by the British astronomer Edwin Hubble (1889–1953). Hubble was then in charge of the newly built "100-inch" (2.54 m) telescope at the Mount Wilson Observatory, which, sitting in the mountains above the (then) small town of Los Angeles, was the largest and best telescope in the world. Using the unprecedented resolving power of the Mount Wilson telescope, Hubble was able to confirm that the faint, small, cloudlike "spiral nebulae" observed in the heavens were, in fact, vast conglomerations of stars like our own galaxy. Moreover, when he measured the spectral lines—highly specific and characteristic wavelengths of light—emitted from the glowing atoms in the nebulae, he found that the light coming from almost all of these galaxies was shifted to the red end of the spectrum (longer wavelengths).

Hubble knew that a possible reason for a systematic red shift was that the galaxies were moving apart; the Austrian physicist Christian Doppler (1803–53) had described how frequency changes with motions of the source (such frequency changes are now known as Doppler shifts). But whereas in the 1920s, when the red shift was first observed, it was not at all clear why the vast majority of the galaxies in the Universe might be flying apart, Gamow and Alpher's theory of a Universe expanding from an initially hot, dense state nicely rationalized this startling observation. As is sometimes the case, however, what we now see as a key theoretical advance was ignored for many years after its publication, perhaps in part because of Gamow's sense of humor. Gamow added the name of his friend and Cornell colleague Hans Bethe (1906–2005) to the paper, because it amused him that the authors' names, Alpher, Bethe, and Gamow, resembled the first three letters of the Greek alphabet.

Gamow realized that if his theory—which the competing theorist Fred Hoyle (1915–2001), in a spectacularly unsuccessful attempt to discredit it, later derisively termed the "Big Bang" hypothesis—was correct, the fiery origins of the Universe should have produced observable consequences in the modern cosmos. Specifically, the contemporary Universe should still be filled with the heat of that originally dense, high-energy state, but the heat would have cooled enormously over the inter-

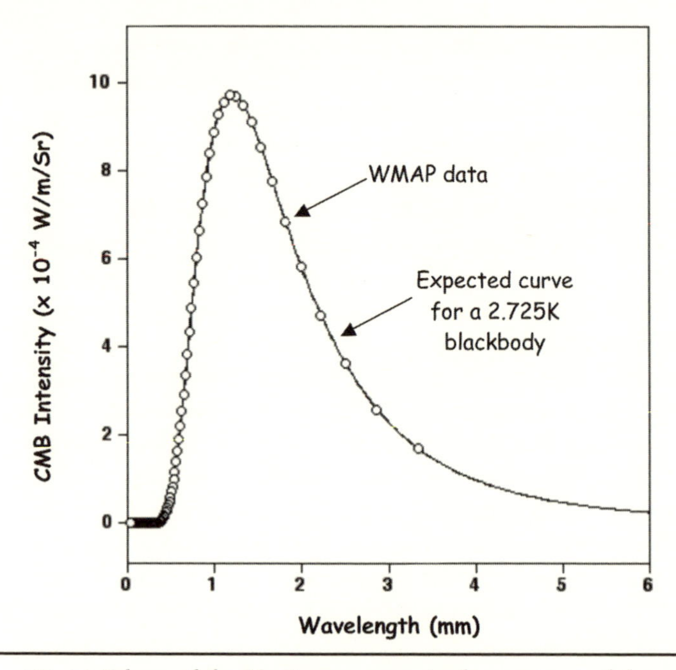

FIG. 2.1. Echoes of the Big Bang are seen in the spectrum of the cosmic microwave background (CMB). The radio-wave photons that this spectrum comprises are the red-shifted, cooled remnants of the hot sea of photons that filled the Universe at the time of recombination (discussed later in this chapter), some 378,000 years after its origins. As shown by the fitted line, the relative intensities of the CMB (measured in W/m/Sr) now exhibit the spectral characteristics of a blackbody (perfect radiator) at a temperature of precisely 2.725K. (WMAP, Wilkinson Microwave Anisotropy Probe.) (Data courtesy NASA/WMAP Science Team)

vening eons. Using his estimate for the age of the Universe, Gamow predicted that the skies would behave as if the Universe were a blackbody at a temperature of approximately 4°C above absolute zero (~4K). With this insight in hand, it was trivial to show that the radio hiss observed from New Jersey corresponded to a blackbody with a temperature of about 5K (since refined to 2.725 ± 0.002 K; fig. 2.1). Rather than the prosaic hiss of pigeon droppings, the physicists Penzias and Wilson had observed the hum of the cooling remnants of the origins of the Universe itself.

Big Bang

Our contemporary understanding of physics is sufficiently advanced that cosmologists have been able to refine the Big Bang model into a detailed description of the origins of the Universe that, some claim, can be pushed back to within 10^{-34} seconds of the origins of time itself. This is all the more impressive when one considers that these events happened 13.7 (give or take 0.2) billion years ago, according to the current best estimates. Here we describe some of the recent discoveries that compellingly support this hypothesis and the implications that the Universe's birth from a Big Bang has for the origins and evolution of life.

From our perspective as astrobiologists, the "interesting bits" started when the Universe was a million, trillion, trillion times older, when it was a relatively ancient millionth of a second old. At this point, everything in the Universe, all of the matter—and energy—in you, in Pluto's moon Charon, and in the most distant star in the heavens, was compacted together in a dense, unimaginably hot plasma estimated to be at a temperature of 10^{13}K (at these sorts of temperatures, the kelvin scale is equivalent to Celsius, °C). At this temperature, the mean energy per photon (remember: temperature is proportional to the mean kinetic energy) is higher than the energy bound up in the mass of a proton or neutron (which can be calculated from the nucleon mass using Einstein's equation $E = mc^2$). This meant that, when two photons collided, they could spontaneously convert into a proton-antiproton or neutron-antineutron pair. Conversely, when proton-antiproton or neutron-antineutron pairs collided, they annihilated one another, producing—you guessed it—two high-energy photons. Before the first millionth of a second, the rate at which neutrons and protons were produced equaled the rate at which they were destroyed.

After this point, no new protons or neutrons were formed, and the existing nucleon-antinucleon pairs were busy annihilating one another. For reasons that remain perhaps one of the greatest mysteries in current cosmology, the "particles" outnumbered the "antiparticles" by one part in a billion, and thus the annihilation did not quite go to completion. Were this not true, there would be no matter in the Universe, so this cosmological mystery has such far-reaching consequences as our very existence.

The electron, and its antiparticle the positron, are more than 1,800 times lighter than a proton. Thus it was not until the Universe was 14 seconds old and at a relatively temperate temperature of 3 billion K that electron-positron pair formation stopped and the total number of electrons (again, for some reason electrons outnumber positrons by one part in a billion) settled down to its current value.

At this point the Universe was made up of a hot, dense sea of electrons and nucleons. Nucleons. Neutrons and protons. But not nuclei. The weak nuclear force holds together neutrons and protons to form

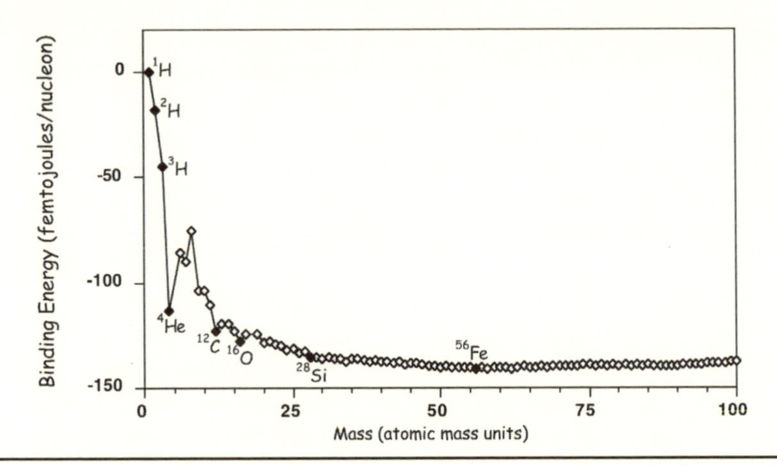

FIG. 2.2. Some nuclei are more stable than others, which explains why stars shine, and which constrained the nuclear reactions that took place during the Big Bang. For example, the nuclear binding energy of ^{4}He (measured in femtojoules, or 10^{-15} J) is significantly greater than that of hydrogen (^{1}H), deuterium (^{2}H), tritium (^{3}H), or helium-3 (^{3}He), and thus fusion of all of these nuclei to form helium-4 (^{4}He) dominated nucleosynthesis during the first few minutes of the Big Bang. The dinuclear fusion of ^{4}He with itself or with any of the lighter nuclei is prohibited by the instability (relative to ^{4}He) of all of the nuclei between the masses of 4 and 12. Fusion reactions to form heavier elements require instead the trinuclear fusion of three ^{4}He to form carbon-12 (^{12}C) in a single reaction. Once this barrier is surmounted, additional dinuclear fusion reactions can continue until iron-56 (^{56}Fe) is synthesized. Further fusion consumes rather than produces energy.

atomic nuclei. And at temperatures above 1 billion K, thermal energies are stronger than the forces that hold nuclei together, and thus any conglomeration of neutrons and protons that might have been transiently formed quickly dissociated under the onslaught of the highly energetic collisions taking place in this hot, unimaginably dense state.

It wasn't until it was about 100 seconds old that the Universe cooled below 1 billion K, the temperature at which the mean thermal energy of its particles (nucleons, electrons, and photons) was sufficiently low that neutrons and protons could come together to form the first composite nucleus: deuterium (an isotope of hydrogen, thus denoted ^{2}H, but also sometimes denoted D). Deuterium is, however, significantly less stable than several higher-weight nuclei and readily converts into these nuclei upon collision. For these reasons, only small amounts of deuterium build up over time.

Similarly tritium (^{3}H, or T, consisting of a proton and two neutrons) and helium-3 (^{3}He, two protons and one neutron) are much less stable than helium-4 (^{4}He, two neutrons, two protons) (fig. 2.2). Thus most of the deuterium, tritium, and ^{3}He formed during the Big Bang were rapidly fused to produce the more stable ^{4}He. (On a related note, hydrogen bombs are really tritium and deuterium bombs, because these nuclei fuse so much more readily than hydrogen.)

And then? And then nothing but more of the same. Between the ages of 3.5 and 5 minutes, the Universe was frenetically converting 20% of primordial hydrogen into ^{4}He, leaving behind only traces of the less stable deuterium, tritium, and ^{3}He. But no significant quantities of any heavier nuclei were produced. Why didn't the Big Bang produce heavier elements in any significant quantity? Inspection of the stabilities of the nuclei (fig. 2.2) indicates why: nuclei composed of 5, 6, 7, or even 8 nucleons are less stable than ^{4}He and are thus not formed in significant quantities. Thus, under the high temperatures present in the Universe during the period of nucleosynthesis, they were destroyed by collisions as rapidly as they were formed. The first nucleon that is stable relative to ^{4}He is carbon-12 (^{12}C). But the formation of ^{12}C requires that *three* ^{4}He simultaneously collide (or, more precisely, that a beryllium-8 nucleus collide with a ^{4}He nucleus during the incredibly brief, 10^{-16} second, lifetime of the former). Thus the formation of ^{12}C is a *third-order* reaction, a reaction whose rate varies with the concentration of reactants *cubed.*

But in addition to cooling, the rapidly aging Universe was *expanding.* By the time appreciable amounts of ^{4}He had formed, the Universe had expanded enough that the ^{4}He concentration was low. Too low, in fact, to allow a third-order reaction to occur at an appreciable rate. Thus, only 3 minutes after the start of the Big Bang, and after a fifth of its initial complement of nucleons had been converted to ^{4}He, primordial nucleosynthesis ground to a halt, leaving mainly protons, free neutrons (which decay into protons with a half-life of $\sim$11 minutes), and ^{4}He. Only small traces of deuterium, tritium (which decays into ^{3}He with a half-life of 12.35 years), and ^{3}He were produced, along with still smaller traces of the heavier nucleus lithium-7.

The ratios of the primordial nuclei provide a stringent test of the Big Bang model of the origins of the universe. The ratios of hydrogen to deuterium, hydrogen to ^{3}He, and hydrogen to lithium-7 are extremely sensitive to the precise density of nucleons in the expanding early Universe. If the density of this matter in the early Universe changed even a little, these ratios would change significantly. Since we do not know, a priori, what the density of the original Universe was, we cannot use Big Bang models to predict what the current ratios should be. But, if we measure these three ratios in the current Universe (and correct for the fact that

SIDEBAR 2.1

The Density of the Universe

The Big Bang model makes specific, quantitative predictions about both primordial nucleosynthesis and the anisotropy of the cosmic microwave background (CMB). Both predictions are sensitive functions of the density of the matter in the early Universe. As we have noted, the fact that both the ratios of isotopes produced in the Big Bang and the anisotropy of the CMB independently point to the same early density provides extremely compelling evidence in favor of the theory. But we've pointedly left out the units in which the density of the Universe is measured.

Astrophysicists could, of course, simply measure density in kg/m^3 (after converting energy to mass). However, they prefer to normalize it to a "critical density" that constitutes a watershed for the fate of the Universe. Density divided by this critical density yields the cosmological parameter Ω. It describes the density of the Universe in terms of whether or not there is enough matter in the Universe to eventually slow and reverse the expansion, leading ultimately to the "Big Crunch." If Ω is greater than 1, the Universe contains enough mass to stop the expansion; this scenario is termed a "closed universe." Conversely, if Ω is less than 1, the Universe will continue expanding (and cooling) forever and is therefore "open." The ratios of the primordial nuclei, which as described in the text are a sensitive measure of the density of baryonic (protons and neutrons) matter in the Universe at an age of about 100 seconds, are consistent with Ω being about 4% of that required to close the Universe. Anisotropies in the CMB provide a completely independent measure of the density of baryonic matter in the early Universe, this time at an age of 378,000 years. Recent density estimates based on the observed anisotropies also produce an Ω of $4.4 \pm 0.4\%$ and an open universe.

A third method of measuring the density of baryonic matter in the Universe is to simply look up and estimate how many stars we can see and their masses. These estimates, which are necessarily very crude, suggest an Ω of only 2% of that required to close the Universe; but given the rather large uncertainty associated with determining the masses of distant galaxies, this result is roughly consistent with the much more precise estimates stemming from isotopic abundances and CMB anisotropies. The close convergence of three entirely independent measures of the density of baryonic matter in the Universe is such compelling evidence in fa-

stars have been converting some of the hydrogen into nonprimordial helium over the intervening 13.7 billion years) and find that all three ratios point to the same density, this provides powerful evidence in favor of the Big Bang hypothesis (see sidebar 2.1). In support of the model, the best current measurements of these ratios are consistent with only a narrowly defined range of densities (fig. 2.3).

vor of the detailed Big Bang hypothesis that the theory is no longer seriously in doubt.

Just because we can measure the baryonic density of the Universe, though, doesn't mean we understand what the Universe is really made of or can predict its fate. The trouble is, the gravity of the baryonic mass that we can see in the galaxies is about one-tenth of what is required to account for the rates with which they rotate. The gravitational pull of some kind of otherwise undetectable matter seems to hold the galaxies together and must push Ω much closer toward 1. The origins of this "dark matter" are a complete mystery. All we know is that it is there and that it cannot be ordinary atomic nuclei (or it would affect all of the parameters described above). Still, some candidates have been put forth, such as neutrinos. These subatomic particles, first postulated by Wolfgang Pauli and long thought to be massless, have recently been shown to possess a small mass. Many other, much more exotic particles have also been postulated to account for this astonishingly large extra mass.

But don't get too comfortable. Just as researchers had accepted that most of our galaxy (and of the Universe at large) is made up of dark, nonnuclear matter about which we know nothing, a second major deficit turned up on the balance sheets. Observations of distant supernovae have confirmed that the expansion of the Universe is *accelerating,* an effect that requires a universal mass/energy content *three times higher* than the combined amounts of ordinary and dark matter. Considering the behavior of the Universe as a whole, the balance sheet now (based on the latest Wilkinson Microwave Anisotropy Probe data) is as follows:

4% ordinary matter
23% dark matter
73% some kind of unknown energy

The unknown energy has been dubbed "dark energy." Together with the dark matter and the tiny bit of baryonic matter, dark energy *pushes Ω very near 1.* At present, our best hope for the resolution of this issue rests on new observations of the rate with which distant supernovae are speeding away, including observations from the orbiting telescope known as the Supernova Acceleration Probe (SNAP). These will tell us more directly whether the expansion rate is accelerating or slowing with time, and thus give us a better feel for which side of one Ω falls on. Until then, we are left with the disturbing conclusion that not only do we not understand the ultimate fate of the Universe (open or closed), but our science can explain only 4% of what our Universe comprises.

Nuclear binding energies are many orders of magnitude stronger than the energies that bind together electrons and nuclei to form atoms.* Thus, while it took only 3 minutes for the Universe to cool suf-

*This is why nuclear bombs are so much more powerful than mere chemical explosives.

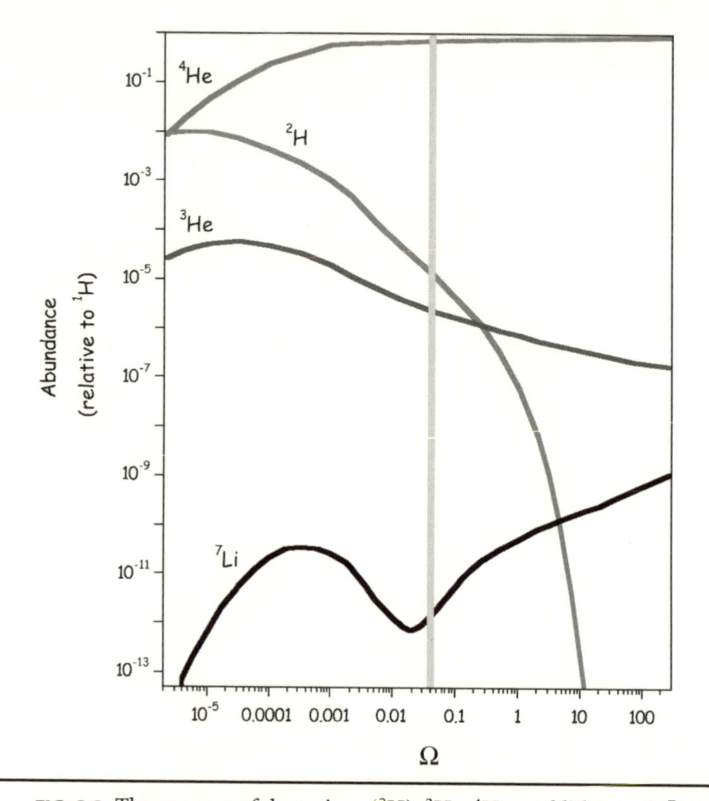

FIG. 2.3. The amount of deuterium (^{2}H), ^{3}He, ^{4}He, and lithium-7 (^{7}Li) (all relative to the amount of ^{1}H) produced during the Big Bang is an extremely sensitive function of the density of the early Universe. The roughly horizontal lines in this figure indicate the relative abundances of the various nuclei expected for a given density (measured in terms of Ω, the critical density needed to produce a closed universe; see sidebar 2.1). Current best estimates of the abundances of these nuclei, all of which fall within the vertical gray bar, are consistent with a density that is about 4% of the critical density required to produce a closed universe.

ficiently for nuclear reactions to freeze, it took much longer to cool to the point where electrons and nuclei could join together to form stable atoms. This "recombination event" (perhaps not the most appropriate of names, given that electrons had never previously been combined with nuclei) occurred when the temperature dropped below about 3,000K, the temperature at which the mean energy of particles in the Universe fell below the binding energy that holds electrons and nuclei together to form atoms. Current estimates are that the Universe cooled to this temperature some 378,000 years after the Big Bang.

The recombination event forever altered the relationship between photons and matter in the Universe. Before recombination, the Universe consisted of plasma; a cloud of bare nuclei and electrons. Photons continuously exchanged kinetic energy with these charged particles and were scattered by them. This interaction equilibrated the photons and the matter in the pre-recombination Universe, such that the photons and the matter were at the same temperature. The scattering would also have made the pre-recombination Universe opaque; plasma, being charged, scatters light in the same way that water droplets in fog cause scattering, and thus the Universe would have been a uniform and brilliant white (if there had been any eyes to see it). Neutral hydrogen and helium, in contrast, are transparent to the visible and infrared photons that filled the early Universe, and thus, after recombination, the scattering stopped. This also stopped the equilibration between matter and the primordial photons, and the two parted company. The matter evolved into galaxies, stars, and us. The primordial photons, in contrast, simply continued to cool with the expansion of the Universe,* until today, some 13.7 billion years later, they have reached the chilly few kelvins that Penzias and Wilson detected.

Fine details of the pervasive cosmic microwave background into which these photons have cooled provide further, extremely compelling evidence in favor of the detailed Big Bang hypothesis. Before recombination, the photons and matter in the Universe were in equilibrium, and thus things that affected matter density affected photon density (temperature). And the early Universe was filled with something that affected the density of matter: sound waves. (Not exactly music, but density fluctuations that obey the same physical laws.) The Universe is finite in size and, before recombination, was relatively small. Under such circumstances, random density fluctuations quickly evolve into standing waves, akin to a violin string responding to random strikes by emitting a pure tone. The wavelength of the standing wave is a function of the density of the medium in which the wave is progressing, and thus if we could measure the wavelengths of the standing waves in the early Universe, we would have a measure of the density of the early Universe. In fact, we can.

The recombination event decoupled the Big Bang photons from the matter in the Universe, and *thus the cosmic microwave background is a*

*They are very much still around, though. They created the hiss that Penzias and Wilson heard and that you can observe at home in your spare time. Or at least you can if you have a television set that's still hooked up to an antenna. If you tune it to a channel that is not broadcasting, about 1% of the screen snow represents relic Big Bang photons red-shifted to television transmission frequencies.

FIG. 2.4. A high-resolution map of the cosmic microwave background shows miniscule, 1 part in 100,000, variations in the 2.725K mean temperature of the background. The angular scale of these fluctuations is a measure of the size—and thus wavelength—of the acoustic waves that filled the Universe during recombination, some 378,000 years after the Big Bang. The size of the oscillations is consistent with a matter density of 4.4 ± 0.4% of that necessary to produce a closed universe (see sidebar 2.1). (Map courtesy of NASA/WMAP Science Team)

fossil of the conditions in the Universe at the time of recombination. If matter in the Universe at recombination was perfectly homogeneous (if it had the same density everywhere), the cosmic microwave background would be isotropic, meaning it would look the same in all directions. If, instead, the matter in the early Universe was filled with standing waves, the cosmic microwave background would exhibit fluctuations corresponding to these waves, and the cosmic microwave background would be anisotropic. In the early 1990s the predicted small (one part in a million) anisotropies were detected by the Cosmic Background Explorer satellite (COBE). Higher-resolution studies of the anisotropy by the follow-on Wilkinson Microwave Anisotropy Probe (WMAP), which was launched in 2001, pinpointed the density within experimental error of the density predicted—completely independently—from the known ratios of hydrogen to the heavier primordial nuclei (fig. 2.4). That two completely independent estimates of the density of the Universe derived from the Big Bang model would converge on precisely the same value provides extremely strong support for the Big Bang hypothesis, and thus the hypothesis is generally accepted as true by the scientific community.

Formation of First Galaxies and Stars; Re-ionization

After about 378,000 years the Universe consisted of relic Big Bang photons zipping through transparent clouds of primordial hydrogen and helium. That wasn't, obviously, a very promising environment for life to arise. Life requires heavier atoms than these two lightweights, and requires seriously concentrated forms of disequilibrium (energy). How did these come about? Initially they arose from the subtle, parts-per-million inhomogeneities produced by the standing wave pattern and reflected in (and much later observed as) anisotropies in the cosmic microwave background.

Despite their parts-per-million level, the acoustically driven inhomogeneities in the early Universe had profound effects on the evolution of the Universe. The slightly denser regions of the early Universe exerted an equally slightly stronger gravitational pull than the rest of the Universe, and thus they began to accumulate even more matter. This new matter increased the magnitude of the originally small fluctuation, accelerating the infall of still more material. Within a few hundred million years (this estimate has been rapidly decreasing from earlier estimates of a billion years as our knowledge of the early Universe improves), the once nearly homogeneous, post–Big Bang cloud of hydrogen and helium had been pulled into a trillion or so lumps, each a few billion times more massive than our Sun. These "protogalaxies" were the seeds of the galaxies we know today; over the next few billion years, protogalaxies would merge to form galaxies. Even today, albeit to a much lesser extent, the process continues, as some of the Hubble telescope's impressive pictures of galactic mergers have shown.

But we have gotten ahead of ourselves. When the Universe was still just a few hundred million years old, it was filled with nothing but neutral hydrogen and helium, contracting here and there to form the first protogalaxies and galaxies. It wasn't until the Universe was nearer a billion years old that the next astonishing thing happened: the first stars were born, shedding fresh light on the Universe, which had descended into darkness as the primordial photons slowly cooled toward microwave energies.

Stars are simply enormous piles of hydrogen and helium, compressed under the weight of their own mass to such densities, pressures, and temperatures (the latter due initially to the kinetic energy associated with all of that mass falling in toward the center) that hydrogen atoms fuse to form helium via the same reactions seen when the Universe was but 3 minutes old. Due to the relatively high density of hydrogen and helium in the early Universe (the Universe had not yet expanded to its current, significantly larger dimensions), the first-generation stars were generally quite massive. Massiveness, in stellar terms, translates to hotter, denser cores, more rapid fusion of hydrogen into helium, greater energy output, and higher surface temperatures. Many of the first stars were prodigiously hot, glowing blue and putting out copious amounts of ultraviolet radiation. This UV radiation allows us to date the formation of the first stars; the light put out by these stars was the first thing since the recombination event that was more energetic than the electron affinity of hydrogen. Thus as the first stars ignited, electrons once again found themselves ripped away from the atoms in which they were bound. Like sunlight burning off a morning fog, this cosmic re-ionization burned off the clouds of neutral hydrogen and helium created by recombination and turned them once again into a plasma of free electrons and nuclei. Using the Hubble Space Telescope to peer at the most distant observable objects, which is the equivalent to looking back 13 billion years to the first 700 million years after the Big Bang, astronomers have recently observed the spectral fingerprints of neutral hydrogen, suggesting that the re-ionization was not yet complete at that time.

The Heavy Elements

The first-generation stars, known for somewhat arbitrary historical reasons as "population III stars," consisted of the elements synthesized in the Big Bang, namely hydrogen and helium. This is not the stuff of which life can be made; life requires chemistry, and the only chemistry that hydrogen and helium can participate in is the formation of H_2. Life is based on heavier atoms, atoms that astronomers (erroneously as far as chemists are concerned) refer to as "metals." So where did these metals, so critical to the origins of life, come from? They were cooked up in stars.

The center of the Sun, to pick our own star as an example, is a toasty 16 million K, a temperature at which the kinetic energy of protons is sufficient to overcome the normal repulsion between two like-charged protons and allow fusion to occur. As we discussed above, the nucleons between ^1H and ^4He are unstable relative to ^4He, and thus the deuterium, tritium, and ^3He that are formed as intermediates are quickly consumed, with the net result being the production of ^4He, two electrons (to balance the charge), and a great deal of energy. This energy, of course, ultimately provides the disequilibrium on which effectively all Terrestrial life is based. It also prevents the Sun from imploding under the weight of its own massive bulk.

Not only is the center of the Sun hot, but it is also under extremely high pressure. This pressure counteracts gravity's incessant attempts to cause the Sun to collapse further, and a "truce" is set up in which the radiation pressure from the core precisely balances the inward pull of gravity. And while this truce will last approximately 10 billion years (for our Sun), it is still temporary; eventually the Sun will consume all of the hydrogen in its core, fusion will slow, and gravity will begin to win. Since the gravitational pull of a larger star is, of course, larger, the counteracting pressure must be higher for stars larger than the Sun. Thus, the equilibrium between the forces that want to tear the star apart and those that want it to collapse generally settle at higher values, meaning higher temperature, density, and turnover of fusion. Because of this, and perhaps counterintuitively, larger stars burn faster and live shorter lives than smaller stars.

And what happens when a star runs out of hydrogen fuel (as will happen to our Sun in ~5 billion years)? As fusion slows, the outward radiation pressure it produces decreases, and gravity begins to win its long tug of war. But as the star contracts, its core temperature rises. When the core reaches 150 million K, the fusion of three ^{4}He to form ^{12}C (and, to a lesser extent, the fusion of ^{4}He with ^{12}C to produce oxygen-16) will commence. The higher temperatures are required to force helium nuclei, which are more highly charged than hydrogen nuclei and thus repel one another more firmly, to fuse. The energy liberated by this fusion reaction causes the star's outer envelope to expand and, since the energy emitted per unit area falls as the star's surface area increases, to cool. The star becomes a red giant.

But why is the next fusion reaction, after hydrogen burning, the burning of *three* ^{4}He to produce one ^{12}C? As we discussed above, none of the nuclei intermediate in mass between ^{4}He and ^{12}C are stable relative to ^{4}He, and thus the formation of these nuclei does not produce energy. In the expanding Big Bang, ^{12}C was not formed, because it requires a trinuclear reaction that is too uncommon at the low densities achieved by the time the Universe was a few minutes old. In fact, even in the highly compressed core of our Sun, the density should be too low for the efficient formation of carbon. The problem is that the vast majority of nuclear collisions are nonproductive. In fact, the first fusion step in hydrogen burning, the fusion of two protons to form deuterium (and an electron), takes place only a few times for every *trillion* collisions. This startlingly poor efficiency occurs because the fusion reaction liberates energy, and this excess energy tears the newly formed nucleus apart, reversing the reaction. The low efficiency isn't much of a problem, though, for a dinuclear reaction; dinuclear collisions occur frequently enough that, even if very few of them are productive, the fusion reaction can occur at a reasonable pace. This is not true for trinuclear reactions, which are rare enough that such a low efficiency should be prohibitive.

Realizing this, the astrophysicist Fred Hoyle (of "Big Bang" fame) postulated that the carbon nucleus must contain a *resonance.* That is, analogous to the electronic excited states of molecules (which are responsible for molecules' absorbing light and thus having color), nucleons in the nucleus exhibit excited states as well. Hoyle reasoned that, given the amount of carbon in the Universe and the improbability of trinuclear collisions, the ^{12}C nucleus must have an excited state that is of precisely the right energy level to funnel off the excess energy that would otherwise cause the nascent nucleus to rupture. This was the first invocation of what is now known as the *anthropic principle:* that is, Hoyle did not necessarily know the precise details of the structure of the carbon nucleus, but because he knew such a resonance had to occur for carbon to be formed and thus, in turn, for us to be here to even ponder this issue, *there must be such a resonance.* Our very existence argues it.

Leaving aside the philosophical underpinnings of the discovery of the carbon nuclear resonance (we deal with the anthropic principle in more detail in sidebar 2.2), we find that, in the far future, the Sun will have shifted from hydrogen burning to helium burning. The increased heat output associated with the switch to ^{4}He fusion will—for reasons that remain unclear—cause the outer layers of the Sun to swell so much that our star will engulf the terrestrial planets. When the Sun becomes a red giant, the Earth will have ended its 10 billion year run. We suppose there is a philosophical point in that, as well, but what to make of it we will leave to the reader.

The ^{4}He in the Sun's core will become depleted after only a couple of billion years of helium fusion, rather significantly less than the 10-billion-year span of hydrogen fusion. The shorter duration of helium fusion occurs for two reasons. The first is that helium fusion produces less energy per nucleon than hydrogen fusion, and thus ^{4}He fusion has to occur more rapidly than the previous rate of hydrogen fusion in order to balance the Sun's gravitational contraction. And, second, because it requires higher temperatures than the fusion of hydrogen, helium fusion is limited to a smaller volume nearer the Sun's center. Thus, when the central core's helium is depleted, the Sun will consist of a central, carbon-rich core, surrounded by a helium-rich shell, surrounded by a thick outer shell of primordial hydrogen and helium. When this happens, our Sun is doomed (bad news for any surviving Earthlings who may have escaped to Europa or Titan when the Earth was destroyed during the onset of the red giant stage a couple of billion years earlier!). It will begin

SIDEBAR 2.2

The Anthropic Principle

A fundamental precept of science since the Renaissance has been the Copernican principle. In 1543, Nicolaus Copernicus (1473–1543) published his *De revolutionibus orbium coelestium libri VI,* which argued that the Earth is not at the center of the Universe—an assertion that revolutionized science. (Indeed, our modern political usage of the word *revolution* stems from the book's title!) When Copernicus's heliocentric model of the Solar System eventually overturned the earlier, Earth-centered universe, all theories based on *exceptionalism,* an assumption that things are different on the Earth than elsewhere in the Universe, became suspect. No longer was man a privileged observer. Even the seemingly benign assumption made by Isaac Newton (1642–1727) that some observers could be said to be at rest in an absolute sense and thus to define a universal reference frame fell when, in 1905, Albert Einstein (1879–1955) published the theory of special relativity.

In astrobiology, however, the Copernican principle itself must be held suspect, for the very fact of our existence has profound consequences, consequences that require we at least partially abandon this long-cherished rule. This idea, called the "anthropic principle," comes in several flavors.

The simplest version is termed the "trivial anthropic principle." Starting from the observation that complex, carbon-based life exists on Earth allows us to deduce that the Universe is more than about a billion years old, because it takes at least a billion years for stars to form, go supernova, and seed the Universe with carbon. Anthropic arguments of this magnitude, while they may at first glance seem trivial (hence the derogatory term employed), have contributed significantly to the history of scientific thought. A classic example comes from the nineteenth century. Based on the known temperature of the Earth and the rate at which a sphere of Earth's volume and specific heat would cool, the British scientist Lord Kelvin (1824–1907) estimated that our planet is only about 10 million years old. Even at that time, however, biologists were in a position to strongly argue that this was insufficient time to allow for the evolution of life's extraordinary diversity. Thus, invoking the trivial anthropic principle (albeit that phrase had not been coined), biologists were able to prove that there was something missing from Kelvin's arguments. What was missing, though, would not be known until the discovery of radioactivity at the start of the twentieth century. We now know that the decay of uranium, thorium, and potassium in the Earth's core have kept it hot for billions of years after the heat of accretion would have been lost.

Our very existence constrains not only the physical parameters possible on our planet but also the range of parameters that describe the Universe as a whole, an idea called the "weak anthropic principle." In this chapter we have already discussed an example of this principle: the fact that carbon-based organisms exist implies that there must be a nuclear resonance in ^{12}C that allows for the efficient production of this nucleus by trinuclear reactions. In 1953, before we knew much about the detailed structure of the nucleus, the British cosmologist Fred Hoyle and his coworkers used this argument to surmise the existence of the ^{12}C nuclear resonance. In a sense, this is also a trivial conclusion; carbon exists, so this resonance must, in turn, also exist. But on reflection, one is struck by the enormity of this coincidence. Were the charge on the proton or the strength of the strong nuclear force to differ by even a fraction of a percent, this resonance would not exist, and neither would life. Does this imply that some higher being must have designed nuclear physics specifically so that life can exist? Not necessarily.

Imagine, for example, some process that produces a large number of universes, each with wildly differing physical parameters and laws. The weak anthropic principle merely states that observers like ourselves will *always find themselves in a universe whose physical properties are consistent with the existence of life.* We should not be surprised that the physical properties of our Universe are consistent with life—*no matter how coincidental this may seem*—because the Universe has to be consistent with the existence of life in order for us to be here to observe that life exists. An important point is that this "observational selection effect" severely limits our knowledge of how probable, or improbable, the formation of life was. We exist, so the probability of life arising in our Universe was not zero. But the probability could be infinitesimally close to zero; because *we have to exist in order to be having this discussion,* we do not know how "lucky" our origins were, only that they did happen. Using the Solar System as an analogy, Martin Rees, an astrophysicist at Cambridge University, describes the importance of this observation by analogy: "If Earth were the only planet in the universe, you'd be astonished that we just happened to be exactly the right distance from the sun to be habitable" (quoted in *Time Magazine,* November 29, 2004). But that absurd improbability becomes much less absurd when taken in context: the Universe almost certainly contains hundreds of trillions of planets. With so many to choose from, it's much less improbable that at least one would be habitable. And the fact that we find ourselves on a habitable planet says nothing at all about how common they are (beyond the fact that at least one exists). This point has such important implications for astrobiology that it bears repeating a third time, in the words of Francis Crick (1916–2004), co-discoverer of the structure of DNA, in his *Life Itself: Its Origin and Nature* (1981): "We cannot decide whether the origin of life on Earth was an extremely unlikely event or almost a certainty, or any possibility in between these two extremes."

While the weak anthropic principle argues that we can say nothing more precise than that the probability of life arising is non-zero, some authors argue that the generation of life is a physical imperative. This "strong anthropic principle" states that the physical properties of this and any other universe must be consistent with the formation of life. Unfortunately it is hard to test such a hypothesis. The entire basis of the weak anthropic principle is the argument that life-free universes cannot be observed, and it is precisely the observation of life-free universes that would be required to disprove the strong anthropic principle.

And how seriously should we take the idea of multiple universes existing simultaneously in some enormously larger hyper-universe? It is perhaps not as far-fetched as you might think; astrophysicists have been moving toward that conclusion, for reasons entirely divorced from the anthropic principle. For example, the most well-respected, detailed model of the Big Bang is called "inflationary cosmology." Inflationary cosmology is founded on the premise that, less than 10^{-27} second after the start of the Big Bang, the Universe went through a brief hyperkinetic period of expansion, swelling from the size of a proton to the size of a grapefruit at a rate millions of times faster than the speed of light (remember: it is space itself that is expanding, not the matter in it, and thus Einstein could make no complaints). Improbable as this may sound, other, more observable predictions of inflationary cosmology have held up to the (admittedly still limited) scientific tests thrown at them. And, it turns out, some variants of inflationary cosmology predict the formation of an infinite number of universes: even after the inflation died down in our region of space, theorists believe, it should have continued (indeed, should still be continuing) in others. And thus, while our part of the hyper-universe pinched off, slowed down, and evolved into the cosmos we see around us, the rest is still spawning an infinite number of new universes—universes that might be ruled by physical laws differing wildly from those we live under. Some superastronomically small fraction of those universes will be habitable, and some still smaller fraction will be inhabited. Inhabited, no doubt, by creatures marveling about the improbability of all the physical laws of their universe being so perfectly tuned to ensure their existence.

to contract again, but our star is insufficiently massive to achieve the core pressures and temperatures required to ignite the further fusion of ^{12}C and ^{16}O. The Sun will thus slowly collapse under its own weight, forming a white dwarf that will cool over billions of years ultimately to become a black, cold ember.

You will have noticed that, for the purpose discussed in this section—namely, how to enrich the Universe with heavier elements—the Sun has been entirely useless, as the elements it produces will be locked up forever in that black ember. We are all made of stardust, but our Sun will have no progeny in this sense. To seed the Universe with atoms that could build living beings, we need much bigger stars. In stars at least eight times heavier than the Sun, the rise in pressure and temperature is sufficient to ignite the fusion of ^{12}C with its own kind and with ^{16}O to form magnesium-24 (^{24}Mg) and silicon-28 (^{28}Si). Given that this reaction requires still higher temperatures and pressures (the larger charges associated with carbon and oxygen nuclei require higher temperatures to overcome the associated larger electrostatic repulsion), it occurs in a still smaller volume of the central core and lasts for an even shorter time than the helium-fusion era. In fact, it lasts only about a thousand years! And when the core becomes depleted of carbon and oxygen? You guessed it. More contraction, higher core temperatures, and fusion to form heavier nuclei, with each fusion reaction lasting for shorter and shorter periods. This process leaves behind concentric layers of hydrogen, helium, carbon, and so on, like some (weirdly spherical) celestial layer cake.

But this cycle does not continue ad infinitum. In the last step, ^{28}Si burns to iron-56 (^{56}Fe), and if you look at the chart of nuclear binding energies (fig. 2.2), you will see that ^{56}Fe is at rock bottom. Further fusion would consume rather than liberate energy. In just hours, the silicon-fusion reaction burns to completion, leaving behind a small, iron-rich core in which no further fusion is possible. Catastrophic collapse, postponed for so long, can be averted no longer.

When energy production in the core stops, the outer layers collapse into the dense, iron-rich core *within seconds.* The infall rebounds off the core, causing a massive shock wave that ricochets through the outer, lighter, *still fusible* layers. This greatly compresses and heats the outer layers, producing a massive pulse of nuclear fusion. The fusion pulse produces an extraordinary density of free neutrons, which avidly combine with any nuclei with which they collide (remember: neutrons are neutral and thus need not overcome the electrostatic repulsion of the nucleus). Thus the neutron pulse generates massive amounts of extremely heavy, neutron-rich isotopes (same atomic number, higher atomic mass). These extremely neutron-rich nuclei are unstable and rapidly decay, typically by the emission of electrons. This converts the excess neutrons into protons, raising the nuclei's atomic number and producing all of the stable nuclei heavier than iron. The rebound of the infalling material from the dense iron core, in turn, spews this new material into space with enormous force. In seconds, a star that was many times as massive as the Sun is torn asunder in a titanic explosion that, for a short time, is brighter than all of the rest of the galaxy: a supernova.

Studies of the elemental and isotopic composition of the Sun indicate that it must be a third-generation (termed population I—again, for historical reasons) star; that is, two generations of stars formed and went supernova before our Sun was formed, in part, from the resulting debris. Population II stars, the second generation of stars formed after the Big Bang, are also extremely common. In fact, spectroscopic measurements, which allow us to assay the metal content of stars by the colors of light they emit, indicate that most stars in the outer half of our galaxy are from this second generation, with membership in this population indicated by the relative paucity of "metallic" atoms heavier than helium. Careful inspection of the spectra of these stars indicates, however, that they do contain some metal, distinguishing them from the first generation of stars; but unlike our third-generation, relatively metal-rich Sun, these stars were born very shortly after the formation of the galaxy. And what of our Sun's grandparents, the population III stars formed from the primordial gas of the Big Bang? Because the Universe was less expanded then, and the hydrogen and helium were denser and promoted the formation of larger stars, the first stars formed were quite large. In fact, computer simulations of the early Universe suggest that typical population III stars were several hundred times more massive than our Sun. Such massive stars would have gone supernova within a few million years and would have rapidly contributed to the metal composition of the second-generation, population II stars. The short lifetimes of the population III stars neatly account for the fact that, while we see many metal-poor population II stars in our galaxy, intensive searches have identified only two candidate stars that are even close to being metal-free (with a metal abundance less than 1/300,000 that of the Sun) and thus may belong to the most ancient group of stars.

Stellar and Galactic Requirements for the Origins and Evolution of Life

So far it seems easy: make a universe, fill it with hydrogen (and helium), let it contract, ignite, and produce some metal-rich third-generation stars, and the stage is set for life, or at least potentially life-bearing solar systems, to arise. But it may not be so simple. Evidence collected over the last few decades suggests that stars of just the right size, composition, stability, and galactic location to support life may be rather rare.

A look at the map of nearby stars quickly reveals that our star is not like most of the others in our neighborhood. Indeed, the Sun seems to have several rather special characteristics. First, the Sun is solitary. Because they are born in densely packed stellar nurseries (more on this in the next chapter), about 85% of all stars occur in multiple star systems in which two or more stars orbit around their common center of mass. Because of the complex gravitational tugs associated with being in orbit around—or even near—two or more stars, stable planetary orbits are almost impossible in such systems. (Planets skating a stable figure-of-eight around two stars exist only in science fiction.) Thus they seem ill suited as potential breeding grounds for life.

Second, the Sun is also a relatively massive star. This fact is often underappreciated because the Sun is more or less in the middle of the range between the most and least massive stars. But larger stars are exponentially less common than smaller stars, and thus the Sun is among the most massive 10% of the stars in our corner of the galactic woods. And the size of the Sun is a critical element of its ability to support the origins and evolution of life. Were the Sun smaller, like the red dwarf stars that outnumber it more than tenfold, it would be so cool and dim that the volume of its "habitable zone" (more on this in the next chapter) would be positively puny. For example, the habitable zone of Barnard's star, a typical red dwarf that, at 6 light-years away, is the second closest stellar system to our own, extends out to only one-twentieth of the distance of Mercury's orbit around our Sun! While we know relatively little about the frequency with which Earth-like planets form, it seems unlikely that such a small habitable volume would harbor one. Equally critically, our Sun is not so massive that it burned out before life had time to arise. Only around 5% of all stars lie in the narrow range between being sufficiently massive to produce a large habitable zone and yet sufficiently small to remain stable for more than a billion years.

Lastly, both the metal content and stability of our star are unique among the hundreds of known third-generation stars in its mass range. The Sun's metal content is about 50% greater than that of other stars of its age and type, and it exhibits only about one-third of the brightness variation of these same brethren. Both of these characteristics may have played a critical role in the development of life on Earth, as both heavy elements and a lack of large, sterilizing stellar flare-ups are probably prerequisites for habitability.

To make matters worse for potential extraterrestrials, the Sun's special nature may not be limited to its lack of a companion, its mass, its composition, and its stability. It may also extend to its galactic location. Recent research suggests that the origins of life are no different than so many other aspects of existence for which the three secrets to success are "location, location, location."

The Sun orbits the center of our galaxy at a distance of 27,700 light-years, taking 225 million years to complete a single revolution. This distance is near perfect in terms of the Sun's ability to support a life-bearing planet. Only slightly nearer to the center, the density of stars climbs very rapidly to densities so great that supernova explosions are frequent enough and close enough to sterilize planets on a timescale that is rapid

SIDEBAR 2.3

Weighing the Probabilities

For life to arise and thrive around a star, that star almost certainly must meet a number of important criteria that we have discussed in this chapter. These include the lack of stellar companions (due to the lack of stable planetary orbits around multiple star systems), being of sufficient size to produce a large habitable zone and yet small enough to burn for billions of years without going supernova, and being a metal-rich, population I star residing in a relatively low-eccentricity orbit permanently within the galactic habitable zone (GHZ). The probability of meeting all three criteria simultaneously is given as the product of the individual probabilities:

$$P_{habitable} = P_{solitary} * P_{size} * P_{GHZ}$$

Our best estimates for these parameters, described in the text, produce a $P_{habitable}$ of 0.00075; that is, about 1 star in 1,300 seems to be suitable. Of course, there are an estimated 100 billion stars in our galaxy, so these odds may not be that bad.

On the other hand, as noted in the text, our Sun is significantly richer in metals and significantly less variable than any other characterized star of similar size. If these criteria are also critical—and, given the role that metals play in planet formation and the serious, for lack of a better word, inconvenience stellar variations produce, they may be—we cannot estimate $P_{habitable}$. We only know that it is greater than zero (we are here, after all, so there is a non-zero probability of there being a habitable planet in our galaxy), but perhaps something much less than this upper limit of 1 in 1,300.

compared with evolution. A location nearer still to the core, and the x-rays produced by the massive, multimillion-Solar-mass black hole thought to reside at the center of our galaxy would fry the complex molecules associated with life. Much farther out than the Sun's orbit, all of the stars are, as mentioned above, metal-poor population II stars. Without metals, planet formation is severely inhibited, and what few planets exist are probably poor substrates for life. In combination, these conditions produce what is known as the galactic habitable zone (GHZ). And whereas the GHZ does comprise a fair fraction of the volume of our galaxy, it is a sparsely populated fraction. Our best estimates are that only approximately 10% of the stars in our galaxy reside within the GHZ. The existence of life on our planet, from simple microorganisms to human beings, is a result of the unique conditions that exist in this zone.

Of course, merely being in the galactic habitable zone right now is only part of the equation. Just as critically, in order to bear life a star must *remain* in the habitable zone for a sufficient length of time. And most stars do not. The eccentricity of the Sun's orbit around the galactic center is extremely small; that is, in lay terms, the Sun's orbit traces out an almost perfect circle. Because of this the Sun—and with it the Earth—remains at a near constant distance from the galactic center, which provides a safe haven from the potentially disruptive effects described above. Such low-eccentricity orbits are, however, relatively rare, and the large majority of the Sun-like stars currently in our neighborhood spend a significant fraction of each galactic orbit far too close to the galactic center for comfort. When taken with the relatively small size of the GHZ, the eccentricity of most Sun-type stars is sufficiently great that, so it is estimated, less than 5% of all stars lie permanently in the life-supporting zone. When taken with the stellar size limits and inappropriateness of multiple star systems, it seems that only a tiny fraction of the stars in our galaxy are well placed to support life (see sidebar 2.3).

Conclusions

Our understanding of the origins of the Universe has advanced to an unprecedented degree in recent years; the Big Bang model is now a quantitative and compellingly confirmed model of how everything within and around us came into being. With this understanding of our origins, though, comes an appreciation of the vast number of things that have to be just right for a universe to be habitable. If the Universe were too dense, it wouldn't have survived long enough to make us. If it were too sparse, galaxies and their associated stars and heavy elements would not have formed. Were there no resonance in the ^{12}C nucleus, there would be no heavy atoms, no chemistry, no life. And on, and on. Worse still, these fundamental properties of the Universe are not the only things that had to be just right in order to create conditions for life. The size, and perhaps metallicity and stability, of the Sun, and its low-eccentricity galactic orbit in the middle of the GHZ, all seem to be critical aspects of the recipe that makes our planet habitable.

And what of Gamow, Alpher, Peebles, Dicke, Wilkinson, Penzias, and Wilson? Gamow died in 1968, just three years after the breathtaking confirmation of his theory. A decade later Penzias and Wilson shared the 1978 Nobel Prize for their discovery (although, perhaps unfairly, they did not share it with either Alpher, Gamow's still-living student, or the Princeton group who were the first to explain the implications of the radio hiss).

Further Reading

The Big Bang. Weinberg, Steven. *The First Three Minutes.* New York: Basic Books, 1993.

Cosmology and the anthropic principle. Rees, Martin. *Just Six Numbers.* New York: Basic Books, 2000.

The galactic habitable zone (GHZ). Lineweaver, C. H., Fenner, Y., and Gibson, B. K. "The galactic habitable zone and the age distribution of complex life in the Milky Way." *Science* 303 (2004): 59–62.

CHAPTER 3

Origins of a Habitable Planet

On a cloudy night in March 1993, Carolyn Shoemaker was peering at some photographs of the heavens that she and her husband, Eugene Shoemaker (1928–97), of the U.S. Geological Survey, and their amateur astronomer friend David Levy had taken a couple of nights earlier from atop Mount Palomar in southern California. The photos, taken an hour apart, were part of a multiyear survey of comets and asteroids. The motivation behind the survey was her husband's: Eugene Shoemaker was the world's preeminent authority on *impacts*—that is, on the effects of meteorite and cometary strikes on the solid surfaces of the Solar System. The survey was his attempt to quantify the number of asteroids and comets that might be expected to cross paths with the planets. But what Eugene Shoemaker had really wanted was to see the effects of an impact with his own eyes. He had long dreamed of capping his career as one of the founders of astrogeology by exploring a fresh impact crater, perhaps in one of his favorite stomping grounds such as the remote deserts of Australia's outback. No significant impact, however, had been witnessed during the entire span of recorded human history.*

The photos Carolyn Shoemaker was viewing revealed a strange, fuzzy streak in the sky. "I don't know what I've got, but it looks like a squashed comet" (quoted by Levy, 2000). The problem was that the fuzzy image on the film looked sort of like a comet, but it did not show the typical bright central core of a comet followed by a diffuse tail. Instead, the bright core of the object was an elongated blob, from which not one but several diffuse tails streamed off into space.

In the hour between the two photographs, the object had moved, a sure sign that it was within the Solar System (the stars are so far away they look as if they are fixed in space). But it seemed to be moving in the same direction and with the same velocity as Jupiter, which was nearby

*Although, on November 30, 1954, a 4-kilogram space rock crashed through a roof and hit Elizabeth Ann Hodges of Sylacauga, Alabama, who was sleeping on her couch at the time. Other than a nasty bruise on her thigh, Mrs. Hodges survived unscathed this, the only known impact of a meteorite on a human.

in the sky and imaged on the same piece of film. Could the strange streak simply be a stray reflection of the bright light of Jupiter? The elongated streak did not quite line up with the overexposed spot of Jupiter, suggesting that, unprecedented as it was, the streak was real. It wasn't until the drive home later that night that Eugene Shoemaker struck on a workable theory. What if the comet didn't just seem to move along the same line of sight as Jupiter, but was actually physically near Jupiter in the three-dimensional vastness of the Solar System? If so, the enormous gravity of this, by far our most massive, planet could have raised such large tides in the weak cometary material as to tear it apart. What they had seen, he surmised, was not a single comet but a train of cometary pieces produced by a far too close passage to the giant planet.

As word of the discovery spread, observations of the newly named comet "Shoemaker-Levy 9" poured in. (The name stems from the fact that this was the ninth periodic comet, a comet that orbits entirely within the inner Solar System, discovered by the team.) With each new observation, the orbit of this celestial oddity became more precisely defined. The first observations confirmed Eugene Shoemaker's hypothesis; namely, the comet had passed within 120,000 kilometers of Jupiter on July 8, 1992, less than twenty months before its discovery. By early April 1993 the data were plentiful enough to nail down the comet's orbit and, to everyone's surprise, it turned out the comet wasn't in orbit around the Sun at all. It seems that this ancient resident of the outermost reaches of the Solar System had been captured into orbit around Jupiter way back in 1929, and the comet was now a moon (a string of moons) of the giant planet. More startling still, by the end of May 1993 enough data had poured in to define the comet's orbit precisely, and it was found that Shoemaker-Levy 9 was not going to survive its next orbit. Its next pass close to the giant planet, slated for July 25, 1994, would be so close that Eugene Shoemaker would see his dream come true—albeit not an impact on Earth, but an impact of the comet his team had discovered onto the largest planet in the Solar System.

Researchers far and wide studied this "impact of the century" using Terrestrial instruments, the Hubble Space Telescope, and the *Galileo* probe, then still en route to its six-year orbital tour around Jupiter. In all, twenty-one impacts were observed as pieces of the comet train slammed into the deep atmosphere of Jupiter (fig. 3.1). Huge dark welts many times the size of the Earth appeared and lasted for weeks. Spectroscopic studies suggest that large fractions of the estimated 1-billion-ton mass that ploughed into the gas giant must have consisted of substances like water (estimated at ~20 million tons), ammonia, and methane, collec-

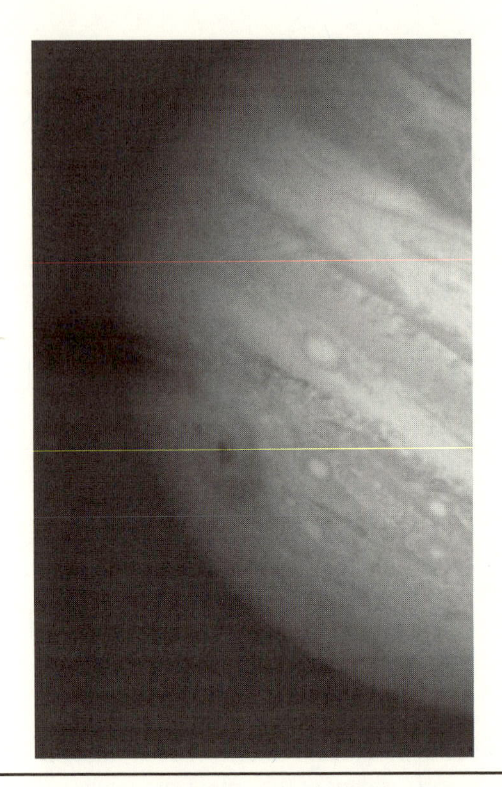

FIG. 3.1. The impacts of the fragments of Shoemaker-Levy 9 left scars in the atmosphere of Jupiter (lower left) that were larger than the Earth and persisted for weeks. The impacts also delivered millions of tons of carbon, oxygen, and nitrogen compounds from the far reaches of space to the gas giant, highlighting the role that Jupiter has played in controlling the dynamics of the Solar System. (Courtesy NASA/STScI)

tively known as "volatiles" in planetary science. And while Jupiter wasn't exactly in need of the extra feeding, the impact of Shoemaker-Levy 9 highlighted Jupiter's huge influence on the movements of stray bodies in the Solar System. With a mass of 0.1% that of the Sun and more than twice that of all the other planets combined, Jupiter's gravitation is a force to be reckoned with.* And as we will see, Jupiter's influence on the motions of objects in orbit around the Sun is so significant that it played a critical role in the evolution of the Earth as a habitable planet.

*According to the Romans, a bearded old man by the name of Iuppiter was the most powerful being in the Universe. He was the god of the clear sky, thunderstorms, and rain; like Big Brother, he saw everything, and was therefore also in charge of law and order.

The Proto-Sun

As described in the previous chapter, the Sun is a typical medium-sized yellow star. From observations of newly born stars in relatively nearby cosmic nurseries and from detailed studies of the composition of the Sun's planets, asteroids, and comets, astronomers have pieced together the story of the birth and evolution of our Solar System.

As the presolar nebula contracted, conservation of angular momentum forced the contracting dust and gas to spiral ever more rapidly around the cloud's center of gravity. Eventually, while the large bulk of the cloud that was to form the Solar System collapsed to form the proto-Sun, a modest proportion of the presolar nebula (theorists disagree about the exact amount, with predictions ranging from a few tenths of a percent to perhaps 2%) fell into orbit around the swelling proto-star. As the cloud slowly collapsed under its internal gravitational pull, its gravitational potential energy was converted into kinetic energy and the inner core of the nebula became quite hot. Eventually, the center of the nebula reached 10 million K, the temperature at which the fusion of hydrogen into helium ignites, and a new star was born. The ignition of fusion occurred before the presolar nebula had entirely collapsed, and thus the early Sun, like all very youthful stars (which are called T-Tauri stars, after a cluster of well-studied examples in the constellation Taurus), was wrapped in a dense cloud of nebular materials (fig. 3.2). The early Sun was quite unlike our present-day star; observations of dozens of relatively nearby T-Tauri stars in the Orion nebula demonstrate that they are much hotter, rotate dozens of times more rapidly, and expel a stream of charged particles (the stellar wind) that is thousands of times more energetic than the solar wind that our now middle-aged star pumps out.

The portion of the presolar nebula (now promoted to the rank of solar nebula) that remained in orbit around the new Sun moved at first with a bewildering array of orbital eccentricities. Nongravitational forces (mainly friction with the nebular gas), though, quickly established a degree of order in the movements and confined most of the material to a thin disk in the Sun's equatorial plane, an arrangement that still holds most of the matter in the Solar System today. How does this work? Just imagine one stray little rock orbiting the Sun on a path that is tilted relative to the thick main disk of gas and dust. During each orbit it would pass through the disk once on its way above the equatorial plane and again half an orbit later on its descent. During each crossing, the stray rock would lose some of its out-of-plane velocity to friction with the gas and dust. Such drag quickly ordered the early solar nebula, herding all of the remaining gas and dust into a thin disk in the equato-

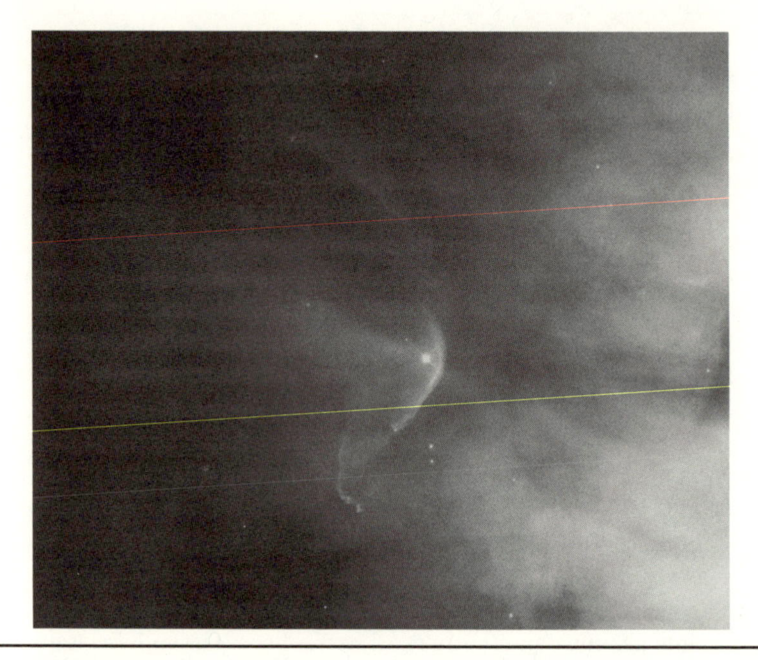

FIG. 3.2. Hot T-Tauri stars, here in the Orion nebula (seen from the Earth as the middle "star" in Orion's sword), quickly begin to blow away the cocoon of gas and dust from which they were born. The solar wind of the T-Tauri star seen here is colliding with the net flow of the nebula, producing a visible shock wave. (Courtesy NASA/STScI)

rial plane of the Sun. While such a disk does not emit visible light, the young star warms it and thus the disk emits infrared radiation. By observing in the infrared, astronomers have imaged such disks of gas and dust around a number of nearby young stars (fig. 3.3).

The Formation of the Planets

The disk of dust and gas that was the early Solar System was far from homogeneous. Heated from within by friction (in effect, the gravitational potential energy liberated as the gas and dust spiraled inward) and from the heat and solar wind of the early Sun, the inner portions of the disk reached thousands of kelvins. Farther out, the temperature of the disk dropped rapidly: dust and gas in the outer reaches of the cloud lost less potential energy and, because the inner cloud was fairly opaque, received far less energy from the Sun. Thus, as the Sun ignited, strong temperature and pressure gradients were set up in the surrounding nebula.

A prominent current theory of the formation of the Solar System is known as the "equilibrium condensation model." The theory was devel-

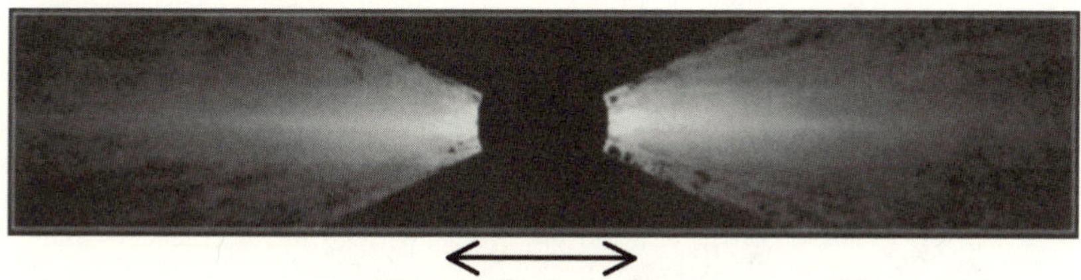

FIG. 3.3. An infrared image of β-Pictoris shows the edge-on disk of gas and dust still present around this relatively young star (the light from the star itself has been blocked out so that it does not overwhelm the much dimmer disk; the diameter of Pluto's orbit is shown for scale). In time the disk will condense into planets. Indeed, this may already have happened: it has been suggested that the distribution of dust, seen by the infrared light it emits, indicates the presence of an asteroid belt and at least one shepherding planet. (Courtesy NASA/STScI)

oped from the physicochemical—and thus compositional—consequences of the temperature gradient. At the extremely low pressures within the solar nebula, liquids were not stable. Thus the only relevant phase change was the sublimation of gases to solids. Astrophysicists tend to call the materials with high sublimation temperatures, such as the iron and nickel that constitute the core of our planet, "refractory." Near the inner reaches of the solar nebula, only the metals and most refractory oxides, such as alumina, condensed to form solid particles. Slightly farther out, less refractory silicates also condensed. At much larger distances from the central proto-star, water, ammonia, and methane (the cosmochemically most abundant molecular forms of oxygen, nitrogen, and carbon, respectively) each in turn condensed to form ices as the temperature dropped with increasing distance.

It is these condensed materials that were the fodder for the production of planets. Conglomerations of particles thus rapidly built up, with dust particles becoming centimeter-sized particles, centimeter-sized conglomerations fusing to form meter-sized boulders, and boulders colliding to form kilometer-sized planetesimals. Due to gas drag, the smaller (centimeter- to meter-sized) particles spiraled inward rather rapidly. As they were thus not in purely Keplerian orbits (i.e., their motions were not solely defined by gravity, but also by friction), they suffered frequent collisions. By the time this *accretion* had collected these into kilometer-sized bodies, an additional force kicked in and accelerated the process: the mutual gravitational attraction of kilometer-sized bodies is sufficiently large that it begins to dominate gas drag and greatly accelerates the rate with which proto-planets grow.

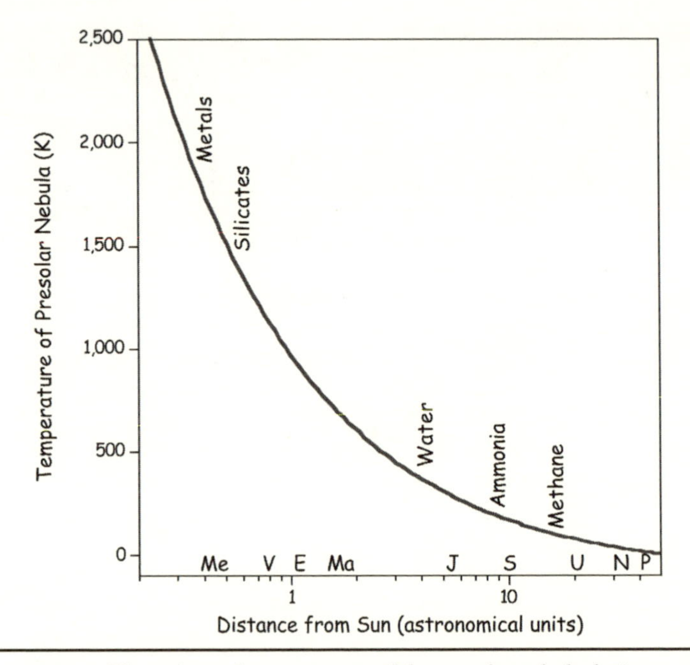

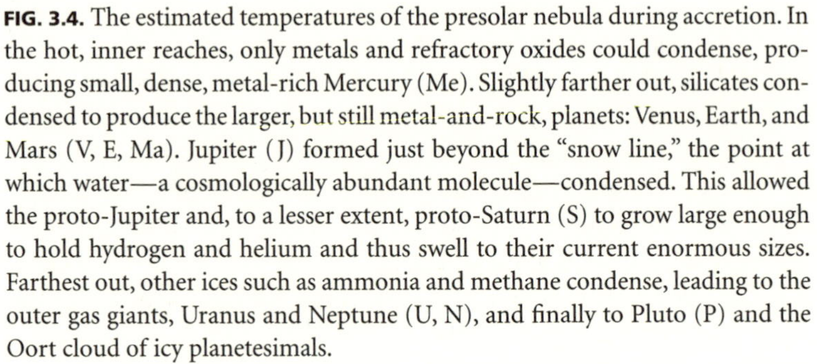
FIG. 3.4. The estimated temperatures of the presolar nebula during accretion. In the hot, inner reaches, only metals and refractory oxides could condense, producing small, dense, metal-rich Mercury (Me). Slightly farther out, silicates condensed to produce the larger, but still metal-and-rock, planets: Venus, Earth, and Mars (V, E, Ma). Jupiter (J) formed just beyond the "snow line," the point at which water—a cosmologically abundant molecule—condensed. This allowed the proto-Jupiter and, to a lesser extent, proto-Saturn (S) to grow large enough to hold hydrogen and helium and thus swell to their current enormous sizes. Farthest out, other ices such as ammonia and methane condense, leading to the outer gas giants, Uranus and Neptune (U, N), and finally to Pluto (P) and the Oort cloud of icy planetesimals.

As described above, the composition of the planetesimals present in the early Solar System varied as a function of how far they formed from the Sun, with metal-rich planetesimals dominating near the center, silicates at the middle distances, and volatile-rich planetesimals farther out in the cold (fig. 3.4). This equilibrium condensation model, so named because the composition of each neighborhood was determined by the equilibrium chemistry that could occur at the temperature found there, roughly predicts the composition of each of the planets. And while the predictions are only rough, because there was some scattering of particles away from where they originated, it explains quite well why the small, solid terrestrial planets, Mercury, Venus, Earth, and Mars, are in

TABLE 3.1
The composition of the Solar System

Planet	Distance from Sun (AU)	Composition	Mass (Earth = 1)	Density (g/cm³)
Mercury	0.39	Metals + alumina	0.06	5.4
Venus	0.72	Metals + alumina + silicates	0.82	5.3
Earth	1.00		1.00	5.5
Mars	1.52		0.11	4.0
Jupiter	5.21	Metals + alumina + silicates + water	317.8	1.3
Saturn	9.58	Metals + alumina + silicates + water + ammonia	95.1	0.7
Uranus	19.28		14.6	1.3
Neptune	30.14		17.2	1.6
Pluto	39.88	Metals + alumina + silicates + water + ammonia + methane	0.003	2.0

Note: AU = astronomical unit.

the inner Solar System; the gas giants, Jupiter, Saturn, Uranus, and Neptune, lie farther out; and farther out still lie Pluto and the other icy, outer worlds of the Kuiper belt (table 3.1).* But while the equilibrium condensation model does a pretty good job, there are a few glaring exceptions among its string of successful predictions: it fails to account for the existence of volatiles on Earth and Venus, or the composition of our Moon.

Near the Sun, the terrestrial planets are formed of various ratios of refractory metals and silicates. As Mercury is the closest to the Sun, it should consist mostly of highly refractory metals, with only a small outer shell of silicate materials. While the chemical composition of Mercury has never been measured directly, flybys of the planet by the *Mariner 10* spacecraft in the early 1970s provided a means of probing the planet's structure and density. According to those measurements, Mercury's exceptionally high density, its gravitational profile, and the occurrence of a strong magnetic field all confirm that the planet's mass is dominated by a dense, metal-rich core. Stepping out from Mercury, we find the other terrestrial planets Venus, Earth, and Mars. These planets formed at a distance from the early Sun at which silicates readily condense and, as indicated by their slightly lower bulk densities, consist of a thick rocky mantle surrounding a metallic core.

Beyond Mars we have the asteroid belt, and beyond that the outer

*The Kuiper belt is a large population of icy bodies, Pluto now being the second largest known. It is named after the late Dutch astronomer Gerard Kuiper (1905-73), who, in 1951, first predicted its existence.

Solar System. The outer Solar System is dominated by Jupiter, which is more than 300 times as massive as the Earth, and the other "gas giant" planets each more than a dozen times as massive as the Earth (itself the largest of the rocky, terrestrial planets of the inner Solar System). A key step toward the acceptance of the equilibrium condensation model was its ability to explain this enormous inequity. Even though we think of Jupiter as a gas giant, it isn't made entirely of gas, and the key to understanding its size and location lies in one of its other components: water.

As oxygen is one of the most cosmically abundant of the elements heavier than helium, water (a combination of oxygen with the Universe's most abundant element) was a major component of the solar nebula. Water, however, is volatile. In the vacuum of space (or, more accurately, the low pressure of the early solar nebula), water vapor does not form liquid water at all, and it did not condense to form ice until the temperature of the nebula decreased to around 150K, at the so-called snow line. Studies of meteorites, which provide ready samples of the asteroids from which they were derived, indicate that asteroids beyond about 5 AU (astronomical units; 1 AU is equal to the mean Earth-Sun distance of ~150 million km) are composed of a reasonable fraction of water. Thus Jupiter, which resides immediately beyond these asteroids, was formed just beyond the distance at which water condensed in the early solar nebula.

Because water must have been abundant in the solar nebula, whose density increased with decreasing distance from the Sun, Jupiter was in a prime location. At this distance from the early Sun, silicates and water ice condensed in larger amounts than anywhere else, and they rapidly accreted to form a rock-and-water planet with a mass ten to fifteen times that of the present-day Earth. Some recent computer simulations of planet formation (which, admittedly, have not yet received widespread acceptance) suggest that this could have occurred quite rapidly—possibly within 5 million years. Once the proto-Jupiter achieved this mass, its gravity became strong enough to pull in *gases.* The terrestrial planets never reached this step; they could accrete only by collisions of solid objects, as their gravity was too weak to hold onto the much more abundant hydrogen and helium that were by far the dominant components of the solar nebula. In contrast, young Jupiter acted like a giant cosmic vacuum cleaner, rapidly sweeping up all the material in or near its orbit. As a result, it swelled over the course of only a few million years and became the largest planet in the Solar System.*

*Some have theorized that the gravitational effects of Jupiter grew so rapidly that, even before the planet had finished growing, its gravity disrupted the accretion process

In the reaches beyond Jupiter, the story is somewhat similar. Saturn rapidly formed a rock-and-water core and began to attract hydrogen and helium. But because the solar nebula was less dense at Saturn's distance (and orbital periods slower—and thus collisions less frequent), Saturn was in a less favorable position than Jupiter and did not accrete as much hydrogen and helium before these gases were driven off by the intense solar winds of the T-Tauri-stage Sun. Uranus and Neptune, although born in still less-dense portions of the nebula, had the advantage of being so far down the temperature gradient that ammonia and methane (which, as the hydrides of nitrogen and carbon, respectively, are also cosmologically abundant molecules) could condense. With these volatiles, they built up rock-and-ice cores of several times the mass of the Earth, but neither planet grew bulky enough to pull in and hold significant amounts of molecular hydrogen or helium before these were blasted from the Solar System. Further out still, the low density of the nebula and long orbital periods lasting for centuries of Earth time greatly slowed accretion, which is why only small, icy bodies were able to form out there. Pluto, for example, is only 1/300 as massive as the Earth.

How rapidly did those condensation and accretion processes occur? Chondritic meteorites have given us a clear indication of this timescale. These rocky space travelers are the most abundant type of meteorite, and they consist of the oldest, most primitive and unaltered material in the Solar System and thus offer a window into the first condensation events. The oldest chondrite precisely dated using radioisotopic methods (see sidebar 3.1) clocks in at 4.559 billion years old. Most others date to within 20 million years of this age, suggesting that the condensation process was very rapid indeed relative to the 4.56 billion year age of the Solar System. Computer simulations of the condensation of the presolar nebula that take into account all of the known physics and chemistry of the problem suggest that, once condensation into meter-sized bodies was complete, the accretion of these planetesimals into the planets we see today required only another 10 to 100 million years. Given the magnitude of the construction process (after all, we are talking about the creation of an entire planetary system here), this is an astonishingly short time. But as we described above, studies of T-Tauri stars indicate that, within a few million years of ignition, the fierce solar wind of the early

---

nearby and prevented the formation of a planet where we now see the asteroid belt. It may also have starved the growing proto-Mars, leaving the Red Planet significantly smaller than it would have been without its giant neighbor.

SIDEBAR 3.1

Radioisotopic Dating

Many nuclei are unstable and decay through a range of processes collectively known as radioactivity. Since the rate of decay of any one type of nucleus (an isotope) is well known and is completely independent of environmental conditions, the decay process can be used as a clock with which to date geological events.

In principle, the concept of radioisotopic dating is straightforward. We simply compare the amount of a given isotope present in a sample today with the original concentration of the isotope. Using the known rate of decay (as given by the isotope's "half-life"), it is a simple matter to calculate how long the decay has been taking place and thus the age of the sample. But radioisotopic dating is straightforward only in principle. The rub is figuring out how much of the isotope was there in the first place.

For some isotopes and some decay routes, the task at hand isn't so hard. For example, the isotope potassium-40 (^{40}K) decays with a half-life of 1.26 billion years by either the emission of an electron to form calcium-40 (^{40}Ca) or the capture of an electron by the nucleus to form argon-40 (^{40}Ar). ^{40}Ca, unfortunately, is the most abundant isotope of calcium and is extremely common in minerals. Thus it is not possible to distinguish the ^{40}Ca produced from the decay of ^{40}K from the ^{40}Ca initially present when the rock was formed, rendering it difficult to use this decay process to date rocks. In contrast, the ^{40}Ar produced by the second type of decay process is well suited for radioisotopic dating. Argon is an extremely inert gas, and thus any ^{40}Ar originally contained in a sample would escape when the rock was molten. Any ^{40}Ar in a rock today, then, must have come from the decay of ^{40}K since the rock crystallized. Thus, by comparing the

Sun would blow away any remaining dust and gas, limiting any further accretion to that involving larger planetesimals. Thus the accretion of the Solar System had to happen rapidly if it was to happen at all.

The Mysteries of the Moon

The accretion model described above makes for a planetary system that changes gradually and logically from the inner toward the outer planets. However, one particular Solar System body falls out of this logic in a number of aspects: the Moon. As we will see, the Moon's gravitational effects here on Earth have exerted a significant influence on life on the planet, so we should have a closer look at why it is there.

Until recently, there was no clear consensus on the origins of our satellite. Before the *Apollo* missions, three theories had been put forth as to the origins of the Moon. The first, the "wife hypothesis," suggested

amount of ^{40}K and ^{40}Ar in a rock, we can readily estimate the time that has passed since the rock formed.

Other isotopes also make suitable targets for radioisotopic dating, but the determination of their original concentrations requires more effort. For example, uranium-238 (^{238}U) transmutes through several steps into lead-206 (^{206}Pb) with a half-life of 4.47 billion years. Some minerals, such as zircon, preferentially exclude lead ions while including the better-fitting uranium ions. Thus, when a zircon crystallizes, its initial lead concentration is zero. Over time, any ^{238}U in the zircon decays into ^{206}Pb. Knowledge of the current ^{238}U content plus the current ^{206}Pb content therefore indicates the original ^{238}U content, which in turn provides a means of dating the crystallization of the material.

Another commonly employed isotope for dating is rubidium-87 (^{87}Rb), which decays into strontium-87 (^{87}Sr) with a 48.8 billion year half-life. Here the availability of an isotope of strontium that is not the product of radioactive decay, strontium-86, provides a means of estimating the original strontium concentration. Combined with knowledge of the current amounts of ^{87}Rb and ^{87}Sr, these numbers allow us to calculate the original and current ^{87}Rb concentrations and thus the time that has elapsed since the rock crystallized.

All three of these radioisotopic clocks—and a handful of others—have been used to date the formation of more than seventy different meteorites and thus provide a lower limit on the age of the Solar System. And while the observed dates vary over the range 4.53–4.58 billion years ago, the best-guess value is 4.56 billion years. Given that many of these meteorites are very primitive (and thus unlikely to have had their radioisotopic clocks "reset" since they condensed from the presolar nebula), this "lower limit" is probably within 10 million years of the actual date on which the presolar nebula started to condense to form the Solar System.

that the Moon had formed elsewhere in the Solar System and only later was captured into orbit around the planet. The "sister hypothesis" was based on the premise that the Moon formed in the same neighborhood as the Earth but failed to accrete with it. The third, not surprisingly called the "daughter hypothesis," postulated that the Moon, perhaps due to the extremely rapid rotation of the early Earth, was torn from the Earth. One imaginative version of the latter theory postulated that the large hole left behind after the split became the Pacific Ocean—this proposal was made before the acceptance of plate tectonics, a theory that implies the Pacific is much younger than the Moon. By the late 1960s, the scientific community had largely discounted the daughter hypothesis, based on the impossibility of the requisite force being generated to pull the Earth apart. But there was no clear advantage for either of the two remaining hypotheses. Everybody expected the question would be settled once the

Apollo missions brought back Lunar samples with which to test the two competing models. It was not.

The *Apollo* missions and the minerals they brought back only uncovered new contradictions. The isotopic distributions found in the Lunar samples (e.g., the ratios of ^{16}O to ^{17}O and ^{18}O in various minerals) are very similar to those found on Earth. This rules out the wife hypothesis, because different portions of the solar nebula exhibit different isotopic patterns. Likewise, however, the sister hypothesis is ruled out by the observation that the elemental makeup of the Moon is vastly different from that of the Earth. For example, rocks from the surface of the Moon are greatly depleted in the more volatile metals such as sodium and potassium. Similarly, the bulk composition of the Moon must be vastly different from that of the Earth: the Moon's mean density is only 3.4 g/cm^3, which nicely matches the 3.3–4.3 g/cm^3 density of pure silicate rocks. In contrast, due to its metal-rich core (iron has a density of 7.9 g/cm^3), the Earth's mean density is 5.5 g/cm^3.

After a couple of decades of hand wringing about these conflicting bits of evidence, the planetary science community has in recent years achieved some consensus regarding the origins of the Moon. According to the current paradigm, the Moon formed via a "daughter-like" mechanism. The force that disrupted the planet, though, was not rapid rotation but instead the impact of a Mars-sized object late in the accretion process that nearly ripped the Earth apart. The force of the impact would have melted both the proto-Earth and the impactor, and splashed an enormous amount of liquid rock into space. Given that the Earth had already differentiated into a dense, iron core and a rocky mantle, the splashed material would have been almost entirely silicates (the impactor was presumably differentiated too, but its metal core would have sunk into and merged with the Earth's). With the heat of the impact, volatiles such as water, and even less volatile substances such as sodium minerals, would have evaporated, leaving the Moon as it is today: a piece of Earth rock (explaining the Earth-like isotopic abundances), depleted of both volatile compounds and metals (accounting for its differing bulk composition). And when did this cosmic collision occur? The absence of erosion on the Moon means that records of even the earliest events in its history remain on its surface today. The oldest rock returned by the *Apollo* missions, known as Genesis, is 4.15 ± 0.20 billion years old. Given that it would have taken a few tens of million years for the liquid splash to accrete and solidify into the Moon, the collision is dated at at least 4.2 billion years ago. Thus the Moon, which, as we will see, is thought to play a vital role in the evolution of life on Earth, was formed in a fluke accident shortly after the formation of the Solar System.

Where Did All This Water Come From?

Next to the strangeness of the Moon, the abundance of water and other volatiles on our planet is the most striking deviation from what the equilibrium condensation model of planet formation would predict. According to this theory, the temperature of the gas and dust that condensed to form the proto-Earth was far too high to let water condense, and thus the Earth should not contain significant amounts of water or any of the other volatile hydrogen, nitrogen, and carbon compounds critical for life. So where did they come from? They came from out beyond the snow line. The effects of massive gas-giant planets are far reaching. In particular, Jupiter's intense gravity affects orbital dynamics throughout the outer Solar System, allowing Jupiter to "clean up" after the late accretion stage. That is, via orbital resonances or close encounters, Jupiter's gravity perturbed the orbits of many of the remaining planetesimals, sending them either into the inner Solar System, where they eventually collided with one of the rocky planets, or into the outer Solar System, where they reside in vast areas of the frozen Kuiper belt and a more distant population of frozen planetesimals called the Oort cloud. This cleanup produced two effects that seemed to be so vital to the formation of life on Earth that, without Jupiter, we would not be here.

The first effect of this Jupiter-induced tidying of the Solar System was to deliver volatiles to the dry proto-Earth; Jupiter's massive gravitational effects perturbed the orbits of icy, volatile-rich planetesimals (asteroids and icy comets) from the outer Solar System and "tossed" them into the inner Solar System, where they collided with—and provided the volatile inventory of—the rocky inner planets. Such delivery is still occurring today. A dramatic example of the probable extraterrestrial origins of Earth's volatile organic inventory occurred on January 18, 2000, when a carbonaceous chondrite meteorite about 5 meters in diameter was seen to explode over Tagish Lake in the far north of Canada's British Columbia province. As the fall occurred in the middle of winter, it was easy to snowshoe out onto the lake and collect fresh, uncontaminated samples of meteoritic material. Fortunately for those of us interested in this sort of thing, one of the few inhabitants of this remote lake was an amateur scientist by the name of Jim Brook who understood the importance of such pristine materials. Venturing out on the lake, he carefully collected samples in clean plastic bags and stored them frozen.*

The couple of hundred meteoritic samples he and later expeditions

*Now you know what to do the next time you see a meteorite land.

recovered are some of the most pristine extraterrestrial material we have on hand. Spectroscopic studies of the Tagish meteorite indicate that it is a good match for the asteroid 368-Haidea, which orbits in the outer reaches of the asteroid belt, beyond the "snow line" at which water is thought to have condensed in the presolar nebula. Sometime in the last 100 million years an impact broke off this chunk of 368-Haidea, and the gravitational perturbations of Jupiter sent it earthward. Of note, the Tagish meteorite is composed of 5% total carbon and about 3% organic material, mostly as aromatic hydrocarbons. The water content of Tagish is somewhat harder to determine, because the meteorite fell on snow. But typical carbonaceous chondrites weigh in at 5%–20% water and, with comets, could have been major suppliers of the Earth's oceans. Even today, Jupiter continues to deliver volatiles to the rocky inner planets.

By also producing a sharp end to the accretion phase of the evolution of the Solar System, Jupiter's cleanup work had a second profound effect on the origins of life on Earth. The history of the end phase of the cleanup is readily visible on any clear, moonlit night: the impact craters left behind by these outer Solar System planetesimals have been preserved on the Moon's face, due to the absence of surface remodeling by processes like erosion or plate tectonics. More quantitative evidence of this cleanup stage was obtained by the *Apollo* astronauts. The 382 kilograms of Lunar samples that the *Apollo* missions brought home allowed geologists to isotopically date various surfaces on the Moon, allowing them to calculate the rate of crater formation during various epochs of the Moon's history. In doing so they found that the cleanup phase ended with what is now called the "late heavy bombardment" some 3.8 billion years ago (fig. 3.5).

One of the last big craters formed in the late heavy bombardment was the Imbrium Basin, which at 1,160 kilometers in diameter is easily visible from Earth as the Man in the Moon's right eye. The size of this crater amply demonstrates that these impacts delivered not only volatiles but also enormous kinetic energy. The planetesimal that formed the Imbrium Basin is estimated to have been about 400 kilometers in diameter. An impactor of this size striking the Earth would provide enough kinetic energy not only to boil all of the water in all of the Earth's oceans, but also to vaporize hundreds of meters of the crust over the entire globe. Because impacts even significantly smaller than the one that formed Imbrium pack sufficient energy to sterilize the entire planet, it is safe to assume that we owe our existence today to Jupiter's having swept the Solar System free of such impactors almost 4 billion years ago.

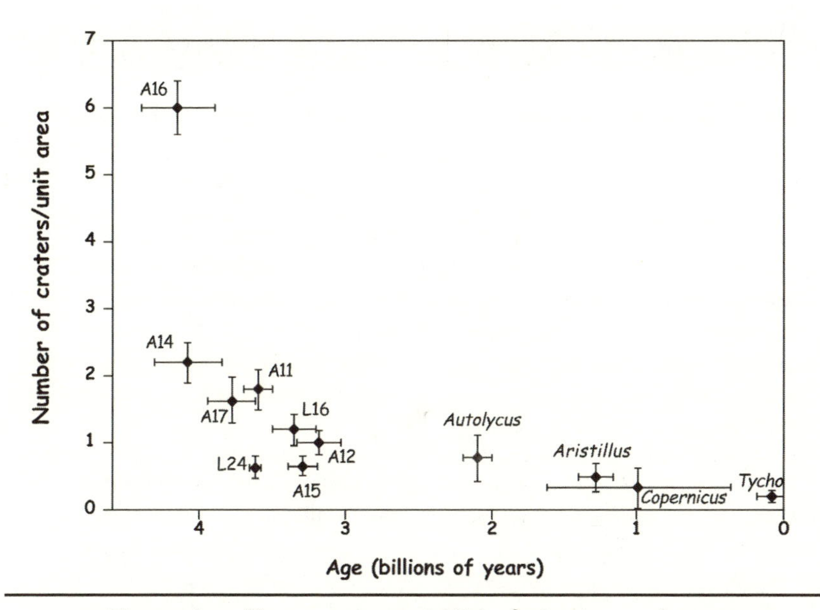

FIG. 3.5. The number of larger craters per 1,000 km^2 of various surfaces on our Moon illustrates the drop-off of the late heavy bombardment at 3.8 billion years ago as Jupiter completed its cleanup of the Solar System. The dates of the various surfaces were determined by isotopic dating of materials brought back by the *Apollo* and *Lunar* missions.

What about Mercury, Venus, the Moon, and Mars?

The other inner, rocky bodies, of course, were also subjected to the late heavy bombardment. But even if you have a rocky planet to start with and get the volatiles delivered to your doorstep, things can still go spectacularly wrong. On innermost Mercury, for example, intense heat and relatively weak gravity allowed most of the volatiles to escape into space. Most, but not all. The axis of Mercury is almost perfectly perpendicular to the planet's orbital plane. Because of this there are no seasons on Mercury, and the bottoms of craters at the poles are forever in shadow and chilled to a few kelvins. These craters thus act as cold traps and appear, as probed by radar studies from the Earth, to be filled with ice! (The *MESSENGER* spacecraft,* now en route to start orbiting Mercury in March 2011, may be able to investigate this question more directly.) The Moon yields a similar picture. While the mean temperature at the Moon's distance from the Sun is much lower than that of Mercury, the Moon's

*Short for *ME*rcury *S*urface *S*pace *EN*vironment, *GE*ochemistry and *R*anging. Phew!

SIDEBAR 3.2

Weighing the Probabilities

So what are the chances of everything coming together just right to form a habitable planet? The question of the frequency with which suitable stars host planetary systems is covered in chapter 9. But this is a reasonable place to ask: what fraction of all the solar systems that contain planets contain planets suitable for the formation and evolution of life? We simply do not know.

One of the issues is how frequently rocky, terrestrial planets form. Again, we discuss this in detail later in the book, but for now let's just point out that current planet-finding techniques are limited to the detection of gas-giant planets (the current limit is something about the size of Neptune), and thus the number of Earth-like planets is difficult to speculate about in any meaningful way. Moreover, we need to know something even more specific: what fraction of planetary systems contain both Earths and Jupiters? The problem is, as noted in this chapter, that without Jupiter to clean up the newly formed Solar System, the late heavy bombardment would have continued effectively unabated, with dinosaur-killing-sized impacts happening once a century or so, and larger, planet-sterilizing impacts happening a few times each billion years.

And if we're interested in the evolution of higher life, a second element comes into play: the Moon. Due to our planet's rotation, centripetal force deforms the Earth from a perfect sphere—the Earth has a distinct bulge at its equator. The Sun's gravity pulls on this bulge and, over the course of millions of years, shifts the Earth's rotational axis. If this force ever succeeded in pushing our planet's axial tilt much below its current value of 23.5°, the seasons would run amok. The Moon, however, weighing in at a quite reasonable 1.25% of the mass of the Earth and residing a mere 384,000 kilometers away, exerts a similarly large force on the Earth's bulge. Because the Moon's pushing and shoving happens over a differ-

gravity is still weaker. The Moon has effectively no atmosphere whatsoever and, save for deep ice-filled craters at its poles too, is completely dry.

But what about Mars and Venus? Mars we will discuss in detail in chapter 9 as a potential abode for life. But Venus, in contrast, does not seem to be even remotely hospitable. This is ironic, because in many respects Earth and Venus are twins. Venus is our nearest planetary neighbor, orbiting only 28% closer to the Sun, and is very similar in mass (Venus is only 18% less massive than Earth). They are also of near identical density, suggesting that, as the equilibrium condensation model would predict, they share similar bulk compositions. But here the similarity ends. Venus is much drier than a bone; the planet holds no significant liquid water, and its atmosphere contains a scant 30 parts per million H_2O as vapor (as compared with 1,000–40,000 parts per million in

ent timeframe from the Sun's pushing and shoving, the Moon prevents the Sun's perturbations from building up and keeps our axial inclination at a nice, mild level. Were it not for this effect, the Earth's climate would fluctuate wildly over the course of a few tens of millions of years (which would not necessarily prevent the formation of life, but could well make it much harder).

Evidence for these putative fluctuations can be seen on our neighboring planet Mars, in which thickly layered terrains near the poles are thought to represent massive climactic fluctuations resulting from the wandering of Mars's axial tilt. Although Mars's current axial tilt is an Earth-like 25.2°, Mars lacks large moons (its moons, Phobos and Deimos, are but a few tens of kilometers across), and therefore the Sun's persistent, small torques build up over time, causing Mars's poles to make massive excursions. Jack Wisdom of the Massachusetts Institute of Technology and Jacques Laskar of the Institut de Mécanique Céleste et de Calcul des Ephémérides and the Paris Observatory have independently shown that, under the influence of these torques, the axial tilt of Mars migrates randomly over tens of millions of years from 10° to a devastating (for climate and the evolution of advanced life) 50°.

So, were it not for our massive Moon, our climate would be unstable and life would be that much less likely to have formed and thrived here. And how improbable was the formation of our Moon? Again, we can only speculate, but the answer may well be "very improbable." The Moon is thought to have formed via an off-center collision between the accreting proto-Earth and a Mars-sized object. The frequency of such collisions is probably low; after all, none of the other three rocky planets have large moons. On the other hand, the icy Pluto has a moon, Charon, that is a whopping 15% as massive as itself. So the probability of such impacts, while low, may not be astronomically low.

Touma, J., and Wisdom, J. "The chaotic obliquity of Mars." *Science* 259 (1993): 1294–97.

our atmosphere). Atmospheric pressure on the surface of Venus is 90 times greater than that on the Earth (given Venus's slightly lower gravity, this implies the Venusian atmosphere is 100 times as massive as the Earth's), and the heavy Venusian atmosphere is 96% carbon dioxide. Carbon dioxide, as is now widely known, is a greenhouse gas. An atmosphere consisting mainly of this gas at 90 times the pressure of our atmosphere creates such an efficient greenhouse that the surface of Venus is kept at an average temperature of more than 460°C,* hot enough to melt lead.

As Venus and the Earth are similar distances from the Sun and have similar sizes and masses, they are roughly equivalently sized "targets" for

*In the shade. And there's no shade.

the delivery of volatiles from the outer Solar System. So why are their current volatile inventories and atmospheric conditions so extremely different? In spite of appearances, Venus and Earth both received roughly the same carbon inventory. Locked inside the Earth's crust are abundant deposits of calcium carbonate—limestone—that are the equivalent of the 90 atmospheres of carbon dioxide that keeps Venus hot. Why is the Earth's CO_2 sequestered safely in rocks, whereas on Venus it remains in the atmosphere, with catastrophic consequences? To answer that question we have to consider that most common volatile: water. In 1979, NASA's *Pioneer Venus* probes entered the Venusian atmosphere carrying a mass spectrometer. One of the primary goals of this instrument was to measure the ratio of hydrogen to deuterium (the heavier stable isotope of hydrogen). In this ratio lies the history of water on Venus.

The velocity of a molecule or atom at a given temperature is inversely proportional to the square root of its mass. Hydrogen, the lightest atom, thus moves on average 40% more rapidly than the average deuterium atom at the same temperature. If a planet is somehow losing hydrogen to space, this enriches the water remaining on the planet in the more slowly moving deuterium. As measured by *Pioneer Venus,* the current H/D ratio on Venus is less than 50:1. And what was the Venusian H/D ratio before the planet lost its water? Jupiter's gravity is so strong that we can be sure it has not lost much, if any, of its original hydrogen. Thus the Jovian H/D ratio provides a measure of the primordial H/D ratio of the Solar System. The *Galileo* atmospheric entry probe also carried a mass spectrometer and, in 1995, directly measured the Jovian H/D ratio at 40,000:1. To have dropped its H/D ratio from 40,000:1 to 50:1, Venus must have lost an amount of hydrogen equivalent to many oceans' worth of water. It is this lost water that is the ultimate cause of the vast climatic differences between the Earth and its erstwhile twin.

Water plays a vital role in the regulation of carbon dioxide levels on Earth, and is thus critical in regulating the Earth's temperature. When atmospheric CO_2 levels rise (due to volcanism, for example), the greenhouse effect increases temperatures, which in turn increases rainfall and the weathering of rocks. This weathering releases calcium and magnesium ions from the rock, which flow into the ocean and there react with carbon dioxide (as carbonate ion) to form calcium and magnesium carbonates—otherwise known as limestone and gypsum, respectively. This sequestration of CO_2 in sedimentary rocks reduces atmospheric CO_2, thus lowering the temperature and slowing weathering. When the CO_2 trapped in the sedimentary rocks is released once again (again via volcanic activity), the process starts anew. This CO_2 cycle forms a negative feedback loop that maintains a constant temperature on the Earth's surface. Without liquid water the cycle fails and CO_2 builds up in the atmosphere, leading to runaway greenhouse warming as, apparently, has happened on Venus.

The carbon dioxide cycle plays an absolutely critical role in, if not the origins of life, at least the maintenance of the Earth's habitability. The issue in question is "the faint early Sun paradox." Stars are relatively simple and predictable systems, and our understanding of their physics is quite mature. It leads us to the well-established extrapolation that early in its life the Sun must have been around 20% dimmer than it is today. (As fusion causes helium to build up in the Sun's core, the zone of fusion moves outward, increasing the Sun's surface temperature and therefore its brightness.) Thus a planet situated at a habitable distance from the early Sun (i.e., a distance at which liquid water could form) would, as the Sun grew brighter, heat up enough to boil its oceans. This problem is exacerbated by the fact that water vapor is a potent greenhouse gas, thus accelerating the heating as the oceans begin to evaporate in earnest. This slowly increasing brightness is a universal feature of stars, and thus astrobiologists need to distinguish between "habitable zones" and "continuously habitable zones." While the former type is the zone in which liquid water can form at a given period in a star's life, the latter is the much, much narrower region in which water stays liquid over billions of years. The continuously habitable zone, while still fairly narrow, is broadened significantly by the carbon dioxide cycle; it is estimated that, for an Earth-type planet in orbit around a Sun-like star, the continuously habitable zone ranges from 0.95 to 1.15 AU. Were the Earth to have formed outside this tight band, life would not have flourished for the billions of years that it has on Earth.*

The importance of the carbon dioxide cycle suggests that plate tectonics—the geological process by which crustal rocks are recycled into the mantle releasing CO_2 (from volcanoes)—plays a critical role in maintaining a habitable planetary environment. The precipitation of carbonate rocks is very efficient and would, over geological time, lead to the removal of effectively all atmospheric carbon dioxide. The resultant reduction in the greenhouse effect would be greatly exacerbated by the fact that snow and ice are white and thus reflective. This, in turn, would lead to further cooling, plunging the planet into a global ice age (termed the "snowball Earth"). Recently reported evidence suggests that this may

*That said, *continuously* is a relative term; the Sun's steadily increasing brightness will push the inner edge of the supposedly continuously habitable zone past us in just two billion years. Don't get too comfortable!

have happened several times in the previous billion or so years of the Earth's history. When the Earth's surface was covered by ice, however, the atmosphere was isolated from the oceans and thus atmospheric CO_2 levels were free to rise. All that was needed was plate tectonics: the carbon dioxide in carbonate rocks is recycled via volcanoes back into the atmosphere. Thus ice-covered oceans lead to increasing atmospheric CO_2 levels, warmer temperatures, the melting of the snowball Earth, and the start anew of the global carbon dioxide cycle.

Liquid water, then, is required to prevent runaway greenhouse warming (à la Venus), and plate tectonics is required to prevent runaway sequestration of carbon dioxide leading to a snowball planet. It is already clear that liquid water is rare in the Solar System. Plate tectonics may also be rare; neither Mercury, nor Venus, nor Mars exhibits any compelling evidence of the effect. Ironically, though, while we are not sure why the Earth exhibits plate tectonics while its near twin Venus does not, it has been theorized that water is the missing ingredient on our nearest neighbor. Without water lubricating them, Venus's crustal plates may be too rigid to subduct.

This, of course, simply pushes the question one step farther: why did Venus lose its water when the Earth clearly did not? The answer to that question resides in the precise locations of the two planets. The mean temperature that a blackbody would achieve were it orbiting the Sun at the same distance as Earth is 255K, which is 18°C below the freezing point of water (were it not for greenhouse warming, the Earth would be a frozen, uninhabitable rock). Because of the low mean temperature at the Earth's orbit, even the middle reaches of the Earth's atmosphere are well below freezing, and any water vapor that may be diffusing up toward the upper atmosphere condenses and falls out as rain or snow (fig. 3.6). The mean temperature at Venus's orbit, in contrast, is above the freezing point, and thus water can diffuse high into the Venusian atmosphere. When water reaches the upper atmosphere it is subjected to the full intensity of the Sun's ultraviolet light, which tears the molecule into its constituent hydrogen and oxygen by photolysis. The oxygen, being extremely reactive, drifts back down and oxidizes Venus's surface rocks (radar images of the surface of Venus indicate the planet is covered with moderately fresh lava flows, and thus there are plenty of fresh, reduced mantle rocks lying around to be oxidized). In contrast, because it is very light, a small but significant fraction of the hydrogen thus produced is moving faster than the Venusian escape velocity and is lost into space. Over the course of the past 4 billion years this mechanism seems to have removed all but a tiny trace of Venus's original inventory of water. In-

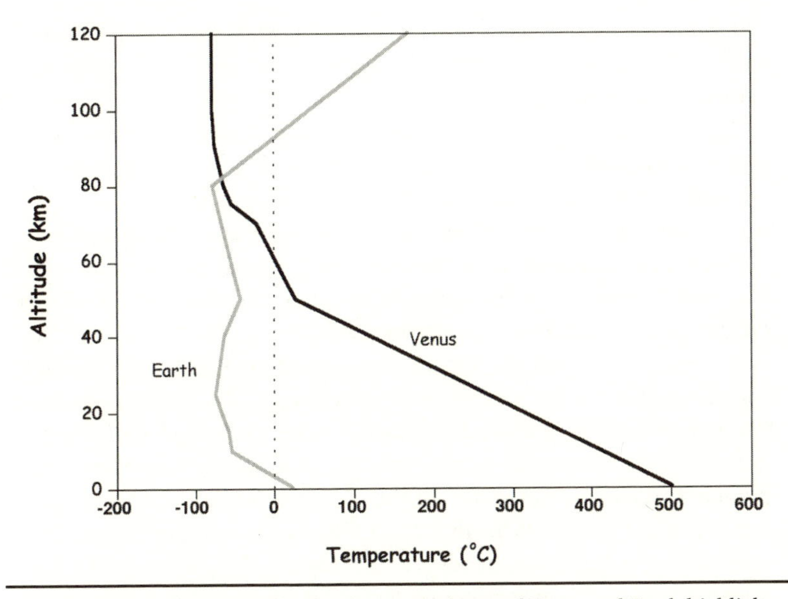

FIG. 3.6. Temperature profiles for the atmospheres of Venus and Earth highlight a key difference between these otherwise very similar planets. The mean temperature of an object at the Earth's distance from the Sun is below the freezing point of water, so the Earth's upper atmosphere is cold enough that water condenses out before it reaches the altitudes at which solar UV can break it into its constituent parts. In the warmer Venusian atmosphere, in contrast, water diffuses to heights in excess of 60 kilometers, where it is photolyzed and its hydrogen lost to space.

deed, it turns out, Earth might also have lost a great deal of water via this mechanism; at 6,000:1, the Earth's H/D ratio is one-seventh that of Jupiter.

Conclusions

Our planet formed in the inner reaches of the solar nebula, and thus like all terrestrial planets is composed predominantly of refractory metals and silicates. These, however, are not the materials of which life is made. Those we owe to Jupiter, whose mighty bulk tossed icy, volatile-rich material from the outer Solar System inward to the nascent Earth. This cosmic cleanup also saved us from later catastrophic, planet-sterilizing impacts, and thus Jupiter played a significant role not only in life's origins, but also by providing the billions of years of stability required for complex life to evolve. But even with these favorable conditions helped along by Jupiter, the development of an Earth-like planet can easily go off toward a state that does not support a diverse biosphere, as the example of Venus clearly illustrates.

The impact of Shoemaker-Levy 9 was just the latest of the final vestiges of this cleanup. It also marked the pinnacle of the long and successful career of Eugene Shoemaker, who, sadly, was killed a few years later in an auto accident while exploring impact craters in Australia's outback. In a fitting tribute to this well-liked and extremely influential planetary scientist, a small amount of his ashes was placed on the *Lunar Prospector* orbiter that was launched not long after his death and that later confirmed suspicions that the deeply shadowed craters on the Lunar poles contain hydrogen, no doubt as water ice. On July 31, 1999, *Lunar Prospector* was intentionally crashed into a crater at the Moon's north pole in an attempt to toss up a plume of dust and steam (alas, though, telescopic observations of the impact from Earth failed to spot the hoped-for water). With this crash, Eugene Shoemaker, one of the most powerful and eloquent advocates of the benefits of astrogeological research, became the first person interred on another world.

Further Reading

The origins and evolution of the Earth. Lunine, Jonathan. *Earth.* Cambridge: Cambridge University Press, 1999.

The improbability of habitable worlds. Ward, Peter, and Brownlee, Donald. *Rare Earth.* New York: Copernicus, 2000.

Eugene Shoemaker biography. Levy, David. *Shoemaker.* Princeton, NJ: Princeton University Press, 2000.

The origins and evolution of solar systems. Lewis, John S. *The Chemistry and Physics of the Solar System.* New York: Academic Press, 1995.

CHAPTER 4

Primordial Soup

In 1952, Stanley Miller was a graduate student working under Harold Urey (1893–1981) at the University of Chicago. Urey, who had won the 1934 Nobel Prize in Chemistry for his pioneering work separating stable isotopes, had become interested in the issues surrounding the origins of the Solar System and the origins of life (the prestige that comes with winning a Nobel generally gives its recipients carte blanche to work on crazy ideas). A question that Miller and Urey particularly wished to address was how the complex molecules around which (at least) Earthling biochemistry is built could have been synthesized from simpler, primordial components before biology itself had begun.

Miller and Urey planned to build on a hypothesis put forward in the early 1920s by the Soviet scientist Aleksandr Oparin (1894–1980) and, slightly later, independently proposed by the Scottish scientist J. B. S. Haldane (1892–1964),* both of whom theorized that life arose via the slow "evolution" of chemical systems of increasing complexity. Speculating on the process, Miller and Urey concluded that the early atmospheres of terrestrial planets are—like that of Jupiter and the other gas-giant planets—filled with methane, ammonia, and hydrogen. Clearly, too, the oceans of a primordial planet would be filled with water still warm from accretion, and thus the skies would be filled with clouds. Miller and Urey speculated that these clouds provided an energy source—lightning—with which to drive the reactions that would be necessary to "activate" the readily available small molecules and link them together into larger, energetically less favorable molecules. Under the energetic activation of lightning, they hypothesized, Oparin's and Haldane's "chemical evolution" would proceed at least to the synthesis of some of the basic building blocks of life.

To test this hypothesis, Miller built an apparatus consisting of two

*Who, oddly enough, was also a communist. Perhaps the antireligious nature of Marxism produces fertile breeding grounds for speculations on the origins of life? Politics aside, Haldane is perhaps best known for Haldane's Law, which states: "The Universe is not only queerer than we suppose, but queerer than we *can* suppose" (in his essay "Possible Worlds," 1927).

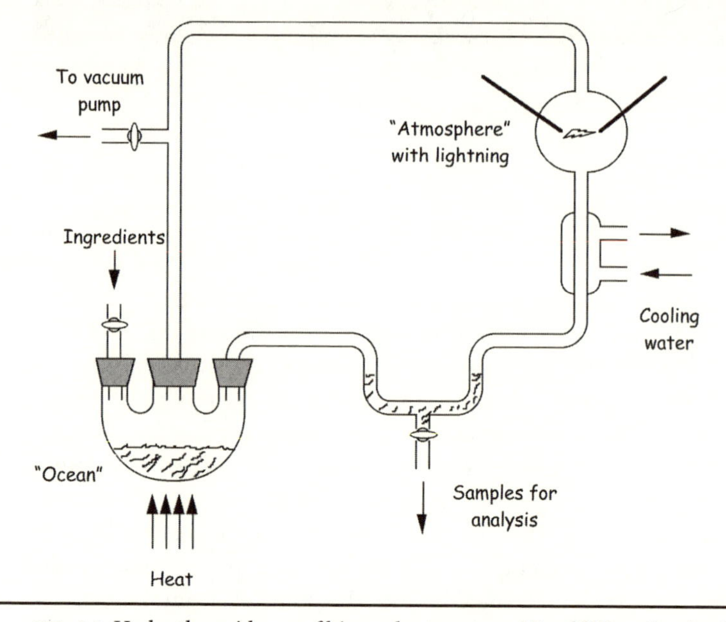

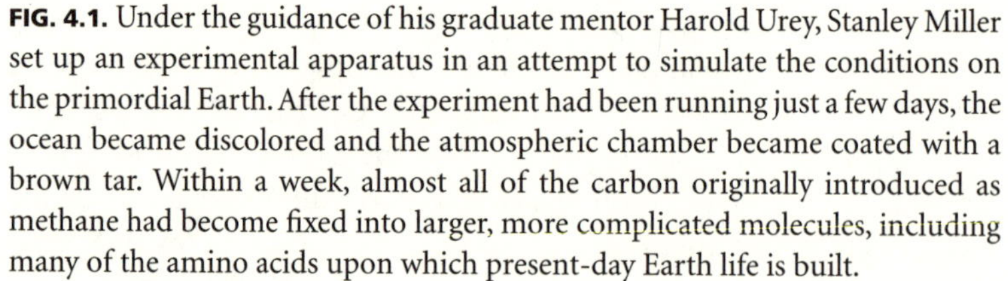
FIG. 4.1. Under the guidance of his graduate mentor Harold Urey, Stanley Miller set up an experimental apparatus in an attempt to simulate the conditions on the primordial Earth. After the experiment had been running just a few days, the ocean became discolored and the atmospheric chamber became coated with a brown tar. Within a week, almost all of the carbon originally introduced as methane had become fixed into larger, more complicated molecules, including many of the amino acids upon which present-day Earth life is built.

connected glass spheres (fig. 4.1). In the lower of the two he placed water as a mimic of the primordial ocean. He filled the upper sphere with methane, ammonia, and hydrogen—Urey's hypothesized primordial atmosphere. Miller placed two electrodes in this "atmosphere," through which he passed a high-voltage discharge mimicking lightning. The connections between the two spheres were such that, when heat was applied to the "ocean," water vapor would rise into the lightning-filled atmosphere. A second tube connected the atmosphere and the ocean by way of a water-cooled condenser, in which water vapor would condense, mimicking rain and setting up a simple hydrological cycle.

Legend has it that Miller sat by his experiment day and night for a week while the simulated lightning crackled and the faux ocean boiled and the vapor condensed as rain. After a few days, the once clear ocean became yellowish and then brown, and the discharge chamber became coated in an increasingly thick tar. After a few more days, Miller removed samples from the ocean for analysis. In a striking confirmation of Urey's

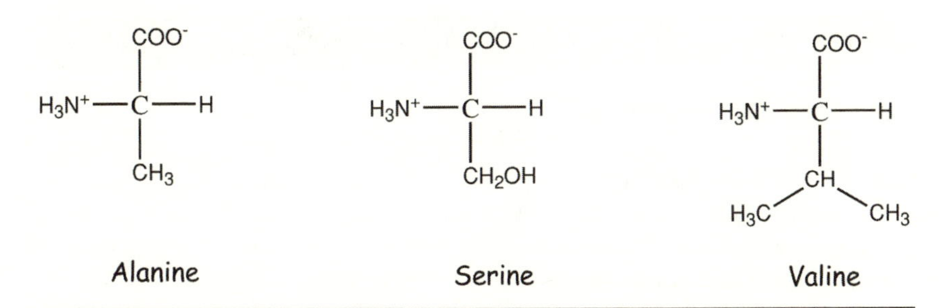

FIG. 4.2. Amino acids are small organic molecules containing a carboxylic acid group ($-COO^-$) and an amino group ($-NH_3^+$). The α-amino acids, three of which are shown here, are a subclass of amino acids in which the amino group is one carbon removed from the carboxylic acid. α-Amino acids play fundamental roles in Terrestrial biochemistry, both as important intermediates in metabolism and as the building blocks of proteins.

TABLE 4.1

Typical Miller-Urey reaction products formed under strongly reducing conditions

Compound	Yield (% total fixed carbon)	Compound	Yield (% total fixed carbon)
Formic acid	4.0	Succinic acid	0.27
Glycine	2.1	Sarcosine	0.25
Glycolic acid	1.9	Iminoaceticpropionic acid	0.13
Alanine	1.7	*N*-methylalanine	0.07
Lactic acid	1.6	Glutamic acid	0.05
β-Alanine	0.76	*N*-methylurea	0.05
Propionic acid	0.66	Urea	0.03
Acetic acid	0.51	Aspartic acid	0.02
Iminodiacetic acid	0.37	α-Aminoisobutyric acid	0.01
α-Hydroxybutyric acid	0.34		
α-Amino-*n*-butyric acid	0.34	Total	15

hypothesis, Miller found the ocean was now a rich soup of higher-molecular-weight carbon compounds. The mixture, which represented 10%–15% of the total carbon input as methane, contained several percent amino acids, the building blocks of proteins (see fig. 4.2 and sidebar 4.1). Under what seemed to be plausible, prebiotic conditions, Miller had shown that even the simple, single-carbon, single-nitrogen, and single-oxygen inputs of methane, ammonia, and water would spontaneously form many of the most fundamental building blocks of life on Earth. (Typical Miller-Urey reaction products are listed in table 4.1.)

Miller's "primordial soup" was hailed on the front page of the *New*

SIDEBAR 4.1

An Earth Biology Primer

An intellectual hurdle that the field of astrobiology faces is its extremely limited data set: we have only a single example of biology to study—as we discuss in chapter 6, every living thing on Earth arose from a single, fairly sophisticated common ancestor—and thus our views on the range of the possible must be somewhat limited. Still, as the only example available for study, Earth life provides our only tangible source of insights, and thus we will constantly return to Earthling biology during our broader discussion of life in the Universe. To do so, however, requires that we are all up to speed with some of the basics of our own chemistry.

Contemporary life on Earth is based on deoxyribonucleic acid (DNA), ribonucleic acid (RNA), and proteins—three types of polymers. DNA serves as the genetic material and contains the information necessary to synthesize the latter two polymers. Proteins, and to a much lesser extent RNA, are the catalysts that perform the myriad of chemical and mechanical functions required for us to thrive and reproduce. They are, in effect, the hands that our genes employ to ensure that they reach the generations that follow. (Like it or not, every aspect of our biology is focused 100% on ensuring that our genes are handed down to our grandchildren.)

DNA is a long, information-containing polymer of nucleotide monomers. Each nucleotide is made up of a sugar (the sugar ribose modified such that it is missing one oxygen atom; thus "deoxyribose") linked to the adjacent monomers in the chain via a phosphate group. Hanging like a pendant from the sugar is one of four nucleobases: adenine (A), guanine (G), cytosine (C), or thymine (T). It is the sequence of these nucleobases in the DNA that encodes the information necessary to make RNA and proteins. The four bases occur in complementary pairs; adenine binds weakly but specifically to thymine, and guanine to cytosine. This nucleobase complementarity allows two complementary DNA polymers to come together to form a double helix. It also provides a means of copying a strand of DNA by using it as a template to make the complementary sequence. RNA is much like DNA, save its being based on ribose rather than deoxyribose (more on this in chapter 6) and containing the nucleobase uracil in place of thymine. Because uracil also forms weak, specific bonds to adenine, DNA can and does serve as a template upon which to build specific complementary sequences of RNA.

Proteins are also polymers, but their monomer subunits are amino acids rather than nucleotides. An amino acid (more specifically, an α-amino acid) is an organic molecule containing an amino group, —NH_2, on the carbon (the so-called α-carbon) adjacent to a carboxylic acid group, —CO_2H. Polymers of amino acids are generated by linking them via these two functionalities to form a "peptide bond" with the structure:

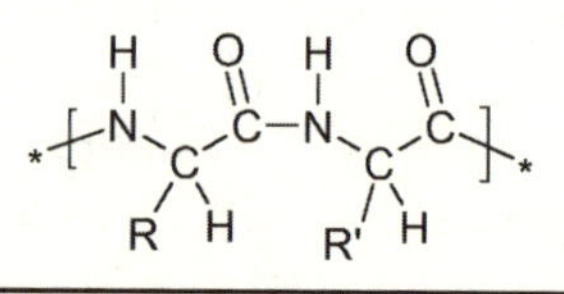

As the two functional groups involved in the polymerization are directly adjacent in the amino acid, the rest of the molecule (here designated by the chemist's shorthand "R," which may consist of just one hydrogen atom or carbon-nitrogen chains and rings) sticks out of the polymer and is therefore called a side chain. Twenty different amino acids, with twenty different side chains and a wide variety of chemistries, are commonly found in proteins on Earth. Under the influence of selective pressures, evolution has come up with billions of different specific sequences of amino acids (about 25,000 different sequences are specified by our genes), each sequence folding into a unique, three-dimensional shape that is defined by these chemistries. Because proteins are, in effect, tools—they catalyze almost all of the chemical and mechanical action in a cell—these three-dimensional structures are critical to their activity; as with all tools, form defines function.

All life on Earth is built around polymers of nucleotides and polymers of amino acids. But did it have to be that way? That is, is it a chemical/physical imperative that life be built using these and only these polymers? Or is the use of proteins, RNA, and DNA simply a historical artifact, a random, early choice of evolution that was present in the last common ancestor of all life on Earth and has since become locked in place? It is, of course, impossible to answer these questions definitively (unless, that is, we find life that doesn't use proteins or RNA!), but that certainly won't stop us from speculating.

It is possible that proteins and RNA are uniquely suited for life in this Universe. For example, α-amino acids are among the most common products of Miller-Urey chemistry. And there's nothing special about Earth in terms of this chemistry; we are safe in assuming that any place in the Universe that contains methane, ammonia, water vapor, and energy will see α-amino acids raining from the skies. Thus, if we believe that life is most likely to arise when it takes advantage of the most common resources available, it is possible that, like you, the snow slugs of Theron Five catalyze their biochemistry with proteins—albeit proteins perhaps built with a set of amino acids that does not overlap entirely with the twenty we Earthlings employ to build our proteins.

York Times as a breakthrough that was sure to lead quickly to a deep understanding of our origins. His result, in some regards the second act of a one-two punch, following the elucidation of the structure of DNA that had been published just a few months earlier, produced tremendous optimism that the "origins" question would soon be answered. Alas, though, while the Miller-Urey experiment provides fascinating and tantalizing insights into possible prebiotic chemistry on the early Earth and elsewhere, it seems that some of this optimism was misplaced. More than half a century later, many fundamental questions regarding prebiotic chemistry—and even more about our origins—remain unresolved.

TABLE 4.2
Volatile "sets" potentially available for young terrestrial planets

Carbon	CH_4	CO, CO_2	CO_2
Nitrogen	NH_3	N_2	N_2
Oxygen	H_2O	H_2O, CO, CO_2	O_2
Hydrogen	H_2, CH_4, NH_3, H_2O	H_2O	H_2O

Volatile Inventory

Miller-Urey chemistry requires an atmosphere and an ocean, materials that are enormously more volatile than the silicates and metals that make up the bulk of terrestrial planets. Thus the origins of the volatile inventory, which we discussed in the previous chapter, play a fundamental role in defining the relevant prebiotic chemistry of a planet. And, as we will see, the chemical form of the volatiles is equally critical to our story.

A quick glance at the cosmological abundances of the elements (see fig. 1.2) and even a cursory knowledge of chemistry suggest some likely candidate volatiles. Hydrogen is, obviously, far and away the most common element in the Universe, but carbon, oxygen, and nitrogen are also reasonably plentiful. Looking over likely (stable) "permutations" of this set of atoms we can come up with a list of the chemically reasonable atmospheres of young terrestrial planets, as shown in table 4.2.

Look at the table closely and you will see that the likely volatile compounds of hydrogen, carbon, nitrogen, and oxygen are grouped into related columns. The relationship by which they are organized in the table is the degree to which the compounds are *oxidized,* that is, the degree to which the various molecules in the atmosphere have given up electrons in a chemical reaction. On the left side of the table are unoxidized (called "reduced" by the chemists) compounds—compounds, such as methane, that are quite willing to give up electrons in chemical reactions. In the center are more oxidized (less reduced) materials, such as nitrogen, which are generally rather unwilling to give up any of their electrons and thus are rather chemically inert. Finally, at the far right we have the equilibrium mixture we would observe for an atmosphere that is so oxidized as to contain free oxygen; oxygen is the second-most *electronegative* atom in the periodic table, which means that, rather than giving up electrons, oxygen takes them from other atoms (only the element fluorine has a stronger affinity for electrons). An atmosphere that is so strongly oxidized as to contain free oxygen is called an *oxic atmosphere,* a term that describes the present, highly reactive atmosphere of our planet. But the oxic nature of our atmosphere is due to the action of algae and higher plants, which use the energy in sunlight to convert water and carbon dioxide to organic molecules and, in the process, produce copious amounts of free oxygen. The question we face in speculating on the origins of life is: how far to the left on this spectrum does the earliest atmosphere of a terrestrial planet typically reside?

The Chemistry of a Newborn Planet

While the Earth provides the best storehouse of information on the evolution of terrestrial planets, we have no direct record of the conditions prevalent during our planet's first half billion years; given Earth's extraordinarily active geology, it is quite rare for Terrestrial rocks to last for more than a few hundred million years at best. Indeed, even the oldest known Terrestrial rocks—relatively small outcroppings in the far north of Canada and slightly larger, more well-characterized outcrops just off the coast of Greenland that clock in at 4.03 and 3.85 billion years, respectively—are more than 500 million years younger than the Earth itself. Thus we are left to speculate on the conditions present for most of the Earth's first billion years and, by extension, on the conditions typically present after accretion of terrestrial planets in general.

Urey was one of the first to consider this question. He reasoned that, while the crust and mantle of the terrestrial planets consist of fairly oxidized silicates, their *bulk* composition is *reducing*. That is, they are primarily composed of molecules that tend to give up electrons in chemical reactions. The majority of the Earth's mass, for example, is contained in its core, which is metallic iron and nickel. Free metals such as these are reduced.* If the differentiation of the Earth into its dense metallic core and rocky mantle and crust occurred slowly enough, the metal, core-forming elements would have equilibrated with the rocky elements in the crust and mantle, reducing them (at the expense of oxidizing some of the iron and nickel). If the crust were reduced, it, in turn, would have reduced the relatively thin layer of oceans and atmosphere; even today, after more than 4 billion years of cooling, the Earth is still geologically active enough to regularly cycle its volatile inventory into and out of the crust (by the subduction of crustal plates in deep sea trenches and the subsequent out-gassing of volcanoes). The Earth's volcanoes put out oceans' and atmospheres' worth of water and gases over relatively short

*In fact, the word *reduced* originates from compounds that could convert (reduce) ores to free metal by providing the oxidized ores with the necessary electrons to liberate the electron-rich metal.

geological times, and thus the crust, oceans, and air are more or less in equilibrium with one another.

In contrast to Urey's speculations, however, geologists have more recently noted that the oldest rocks on Earth seem to have been deposited in a relatively oxidized (but definitely not oxic) environment. This, they argue, implies that the dense metallic iron and nickel sank to form the core too quickly for chemical equilibration with the mantle to occur, and thus the Earth's mantle and crust could have been, as they are today, made up of relatively oxidized silicates. Were this the case, volcanic outgassing would have led to a more oxidized atmosphere. Without a good rock record of the first half billion years, though, we cannot tell which of these two scenarios came to pass. But, as we discussed in the previous chapter (with respect to Venus's losing the equivalent of oceans' worth of hydrogen), there are abiological processes that tend to oxidize atmospheres and thus the crust and mantle of a geologically active planet over hundreds of millions of years. Thus, while the earliest rock record suggests an atmospheric composition nearer the middle of the spectrum from reduced to oxic, this does not prove that, a half billion years earlier, the primordial atmosphere was equivalently oxidized. The ambiguity about the initial atmospheric conditions of the Earth—and thus, by analogy, all terrestrial planets—is a godsend for current theories of the origins of life, as it seems that the prebiotic formation of the likely precursors of life can occur only under reducing conditions. Thus if the early atmospheres of terrestrial planets are relatively oxidized, then the source of the reduced organic compounds thought necessary for the formation of life becomes a potentially significant theoretical hurdle.

Miller-Urey Chemistry

Miller assumed a reduced atmosphere for his experiments—remember: his mentor, Urey, was an early proponent of the equilibrated model of the early Earth and its prediction of a reduced atmosphere. In the decades since Miller's work, his experiment has been repeated many times with many different atmospheric compositions and with energy sources ranging from Miller's simulated lightning to hot rocks (simulated volcanism), ultraviolet irradiation (simulated sunlight), and high-energy subatomic particles (simulating radioactive minerals and cosmic rays). As long as the experimental atmosphere is fairly reducing, all of these variations produce about the same results: namely, a soup of amino acids and other small, simple organic compounds corresponding to about 10%–15% of the total carbon input into the experiment, with the remaining carbon primarily ending up in a complex, high-molecular-weight tar lining the apparatus. With regard to the key question of

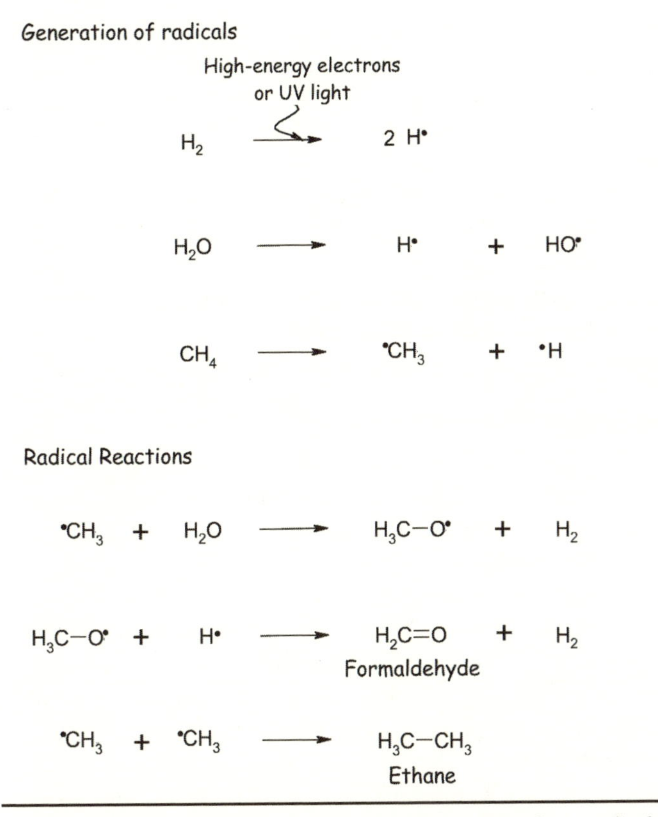

FIG. 4.3. Miller-Urey chemistry is thought to proceed via radical chemistries similar to those taking place in flames. High-energy electrons in the spark discharge (which is thought to mimic lightning) strike atmospheric components, knocking an electron out of them to form "radicals." The unpaired electron left behind is highly reactive and quickly bonds to other atmospheric components to produce higher-molecular-weight organic molecules.

the reduced nature of early terrestrial atmospheres, however, it should be noted that, unless methane and ammonia, or methane, nitrogen, and molecular hydrogen, are present, Miller-type experiments produce relatively little in the way of biologically relevant small molecules.

Miller-Urey chemistry requires an energy source because amino acids and other, more complex organic compounds are unstable relative to methane, ammonia, and water, and thus energy must be supplied in order to synthesize them. In Miller's experiment, the energy source was the lightning-like electric discharge. High-energy electrons in the discharge (or from radioactivity, or high-energy UV photons) tear electrons out of the neutral gas molecules, creating highly reactive species such as methyl and hydroxyl radicals. The hydroxyl radical can attack methane to form two of the simplest oxidized carbon compounds, methanol (CH_3OH) and formaldehyde ($H_2C{=}O$). The methyl radical can attack ammonia, ultimately to form hydrogen cyanide (HCN), or another carbon radical to produce a longer, more complex carbon compound. So long as they remain in the gaseous atmosphere, these more complex molecules also fall prey to the electric discharge, forming still more complex radicals and still longer carbon compounds. A smattering of the radical reactions thought to occur in the discharge is shown in figure 4.3; similar radical chain reactions take place in flames and are a central feature of the chemistry of combustion.

Miller-Urey Synthesis of Amino Acids

Hints about the mechanism by which Miller-Urey chemistry forms amino acids from simpler precursors can be obtained by monitoring the concentration of various species as the reaction proceeds (fig. 4.4). For example, the concentrations of both the simplest nitrogen-containing organic molecule, hydrogen cyanide, and the simple oxidized carbon compounds called aldehydes rise for the first twenty-four hours, before falling off after approximately three days. The reactant ammonia, in contrast, falls steadily as the reaction proceeds, and the concentration of amino acids reaches a plateau after about four days. These trends suggest that hydrogen cyanide and simple aldehydes are intermediates in a reaction that, in net, fixes ammonia into amino acids. A century before Miller's experiment, the German organic chemist Adolph Strecker (1822–71) had reported a synthetic route to the formation of amino acids that similarly employed hydrogen cyanide, ammonia, and aldehydes (fig. 4.5). Miller-Urey chemistry is now thought to be a variant of this "Strecker synthesis," in which ammonia reacts with an aldehyde to produce an imine, a molecule containing a nitrogen-carbon double bond. The carbon in this bond is relatively susceptible to attacks from reactive species such as cyanide, with which it forms an α-aminocyanonitrile. The cyanonitrile group is prone to hydrolysis, in which the addition of two molecules of water and loss of ammonia convert it into a carboxylic acid group, forming an amino acid in the process.

Around 4% of the total carbon input (as methane) into a typical Miller-Urey experiment is converted into amino acids, including ten of the amino acids employed by life on Earth to make proteins (these "proteogenic" amino acids are indicated in bold in table 4.3; fig. 4.2 shows the structure of several of these). Missing from the list, however, are the other half of the twenty proteogenic amino acids used in Terrestrial biochemistry (for those of you who are keeping track, they are: the aromatic

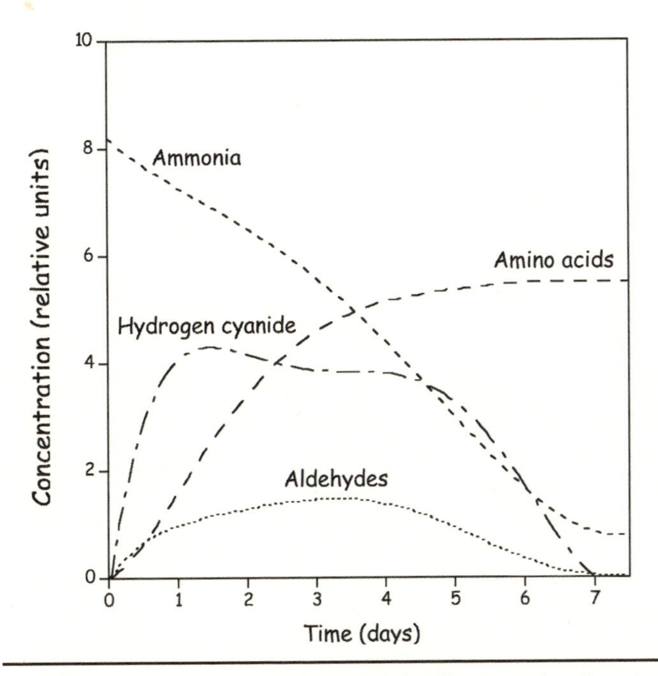

FIG. 4.4. Insight into the mechanisms of Miller-Urey chemistry is provided by the time course of the consumption of NH_3 and the formation of hydrogen cyanide (HCN), aldehydes, and amino acids. The peak in the concentration of HCN and aldehydes at intermediate times suggests these molecules are way-points in a multistep reaction that converts small-molecule precursors into more biologically relevant, higher-molecular-weight amino acids.

amino acids, phenyalanine, tyrosine, and tryptophan; the positively charged amino acids, histidine, lysine, and arginine; the sulfur-containing amino acids, cysteine and methionine; and the amide-containing amino acids, asparagine and glutamine). Since Miller did not include any sulfur compounds in his reaction mixture, the lack of the two sulfur-containing amino acids is not surprising; when H_2S is included in the reaction mixture both methionine and cysteine are produced with reasonable yields. The prebiotic synthesis of the other eight proteogenic amino acids, though, seems to represent a more fundamental problem. These amino acids were almost certainly introduced by *biochemistry* after the origins of life. For the aromatic amino acids, the reason may be the difficulty in synthesizing the aromatic ring via gas-phase chemistry. And instability may be the issue for the amino acids containing amide functional groups ($H_2N{-}C{=}O$); the amide-containing glutamine, for example, undergoes an internal chemical attack that releases ammonia,

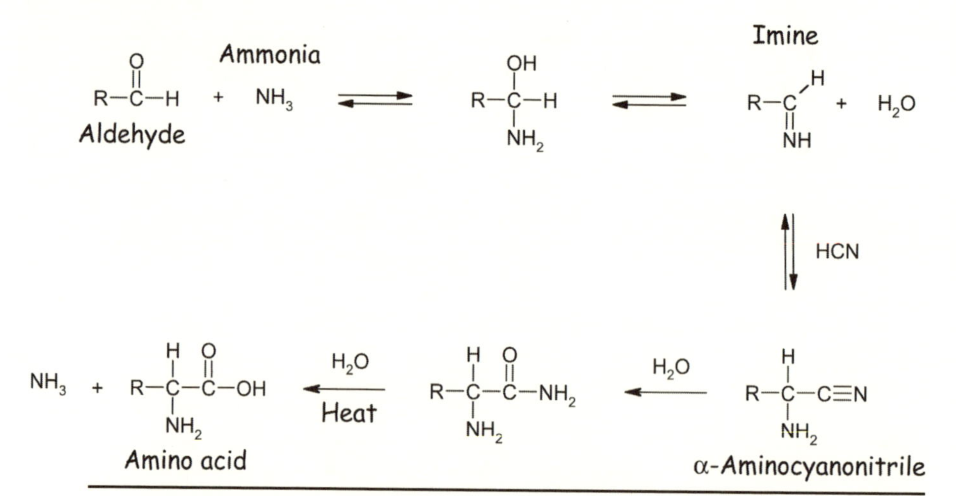

FIG. 4.5. The Strecker synthesis is thought to parallel the mechanism by which amino acids are synthesized in the Miller-Urey experiment, and perhaps to be the source of the amino acids delivered to Earth by the carbonaceous chondrites (discussed later in the text).

TABLE 4.3
Yields of the α-amino acids in the Miller-Urey experiment

Amino acid	Yield (μM)	Amino acid	Yield (μM)
Glycine	440	Norleucine	6
Alanine	790	**Isoleucine**	5
α-Aminobutyric acid	270	**Serine**	5
Norvaline	61	Alloisoleucine	5
Aspartic acid	34	Isovaline	5
α-Aminoisobutyric acid	30	**Proline**	2
Valine	20	**Threonine**	1
Leucine	11	Allothreonine	1
Glutamic acid	8	Tertleucine	0.02

Note: Proteogenic amino acids in bold type.

destroying the amino acid. Given that the half-life of this reaction is only about 100 years, it is difficult to imagine that significant quantities of glutamine could build up over geological time on a prebiotic planet.

Prebiotic Synthesis of the Nucleobases

While Miller-Urey chemistry seems to form plenty of amino acids, from which our proteins are formed, it does not produce significant amounts of the purine (two-ring) and pyrimidine (one-ring) nucleobases from which, in part, our genetic material is synthesized. But in 1961, John "Juan" Oro (1923–2004) found that a mixture of hydrogen cyanide and ammonia in an aqueous solution could be coaxed into forming not only amino acids but also small amounts of the purine nucleobase adenine. Small amounts of hydrogen cyanide are formed early in the Miller-Urey reaction by the impact of electrons on methane and ammonia, and the compound can also be produced by photolysis of N_2 in the presence of methane by UV light high in the atmosphere. Hydrogen cyanide is a fairly high-energy molecule and thus contains the energy necessary to drive a multistep prebiotic chemical pathway. Nevertheless, because four molecules of the compound are condensed in the first steps of the proposed adenine synthetic reaction, relatively high concentrations of hydrogen cyanide are required. How such high concentrations might form is unclear, because hydrogen cyanide is more volatile than water and thus cannot be concentrated by evaporation. As a possible solution to the problem, Miller and Oro proposed that the requisite high concentrations can be achieved by freezing a solution of hydrogen cyanide, which would drive the hydrogen cyanide into the voids between ice crystals. If true, this suggests that prebiotic nucleotide synthesis may have happened only in the polar regions, or on icy planetesimals in the outer Solar System.

The abiological synthesis of the purine nucleobases has been explored in some detail. The process is thought to involve the multistep condensation of four molecules of hydrogen cyanide to form diaminomaleonitrile (fig. 4.6). Under the influence of ultraviolet light (sunlight is sufficient), this compound rearranges and reacts with yet another molecule of hydrogen cyanide to produce the nucleobase adenine in about 7% overall yield. Alternatively, four molecules of hydrogen cyanide can react with the salt ammonium formate to produce adenine with a yield of better than 90%. To achieve this yield, however, the reaction mixture must be heated to dryness: the reaction occurs with the liberation of two water molecules, and thus driving off the water as vapor favors the formation of the nucleobase. A small variation in the first of these synthetic paths to adenine produces a second purine, hypoxanthine, albeit under only a narrow range of conditions and with a miserly 3% or so yield. While the yields of these reactions are not great (and the reactions require relatively complex, potentially rare sets of conditions), Miller has shown that both purines, adenine and hypoxanthine, are stable and thus could have accumulated over geological periods.

The prebiotic synthesis of the pyrimidine nucleobases may have started with the precursor cyanoacetylene, a minor product of the reac-

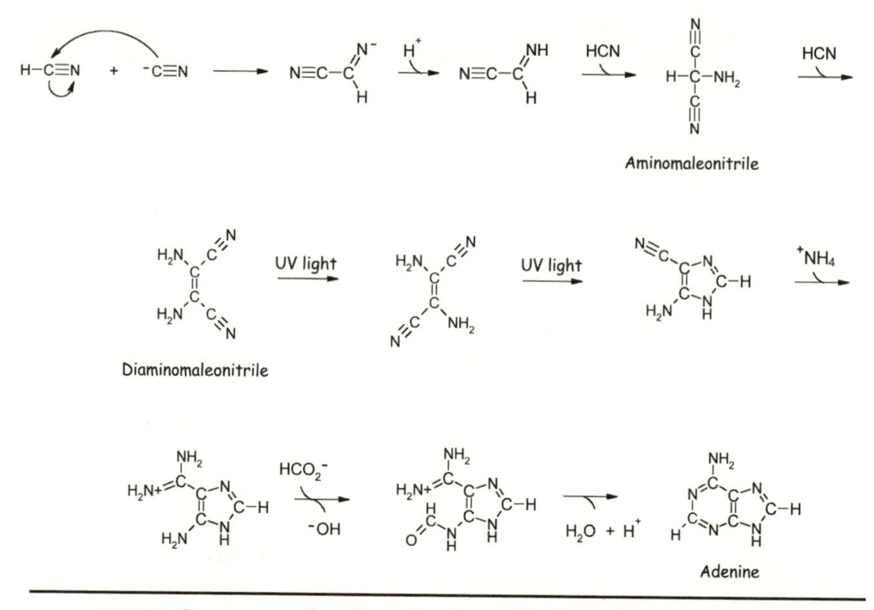

FIG. 4.6. One proposed pathway for the prebiotic synthesis of the nucleobase adenine (a purine) from cyanide (HCN), which is produced in Miller-Urey reactions, through the intermediate diaminomaleonitrile.

tions induced in methane-nitrogen mixtures under the influence of a spark discharge (fig. 4.7). This precursor reacts with water to form cyanoacetaldehyde, and this in turn can react with the small, nitrogen-containing organic compound urea, which is also formed via Miller-Urey chemistry. The intermediate thus formed spontaneously rearranges to form the more stable cytosine, one of the two pyrimidine bases present in RNA. As nucleobases go, however, cytosine is relatively unstable; its half-life in water is estimated to be only around a century at room temperature. The good news about this reactivity is that the hydrolysis product of cytosine is uracil, the other pyrimidine in RNA, thus accounting for the prebiotic synthesis of this critical base. The bad news, however, is that the short half-life of cytosine means it would have been difficult for significant quantities of the nucleobase to build up over geological periods.

While we can point to plausible mechanisms by which several nucleobases are formed, some serious holes remain in our knowledge of prebiotic nucleobase chemistry. For example, adenine, cytosine, and uracil are all components of Earthling DNA and RNA, but hypoxanthine is not. Instead, Earth life uses the purine guanine, and plausible prebi-

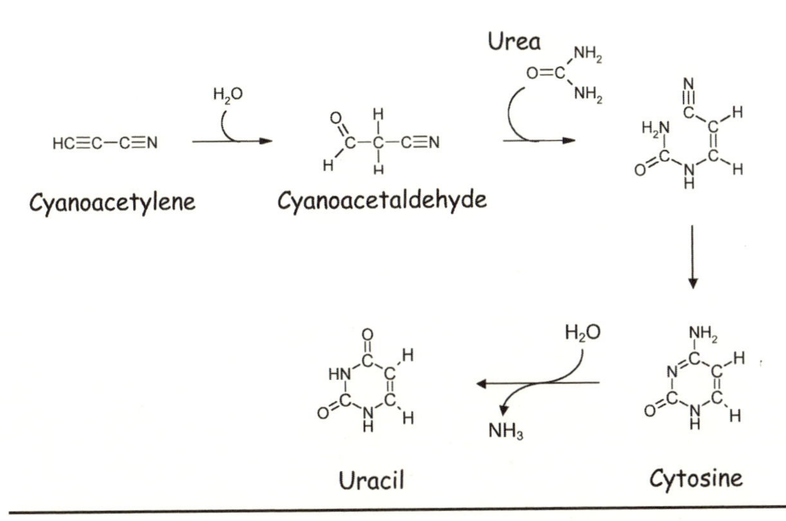

FIG. 4.7. The proposed prebiotic synthesis of the nucleobase cytosine (a pyrimidine) from cyanide (HCN) and cyanoacetylene, both of which are produced in Miller-Urey reactions. Cytosine, in turn, hydrolyzes with the loss of ammonia to form uracil.

otic routes to the synthesis of this important molecule have been harder to come by. In this regard, Oro and Miller suggested a possibility; they reported that guanine can be synthesized via the polymerization of ammonium cyanide (which is formed by dissolving both ammonia and hydrogen cyanide in water) (fig. 4.8). The synthesis works effectively over a broad range of temperatures, but even at very high ammonium cyanide concentrations the overall yield is less than 1%.

The Missing Ingredients: Sugars and Fats

Of course, on Earth life is not built on amino acids and nucleobases alone; the dry weight of a bacterial cell, for example, is about 6% lipids (fatty, water-hating molecules) and 5% sugars and sugar polymers, including the sugars of the (deoxy)ribose backbone of DNA and RNA. Neither lipids nor carbohydrates, however, are readily produced via Miller-Urey chemistry, and yet it seems likely that these materials played key roles in the origins of life. How, then, were appreciable amounts of these materials synthesized under prebiotic conditions? The answer to this question also remains unknown, but several seemingly plausible theories have been put forth.

Sugars are also known as "carbohydrates" because their elemental composition is that of a carbon atom combined with a water molecule:

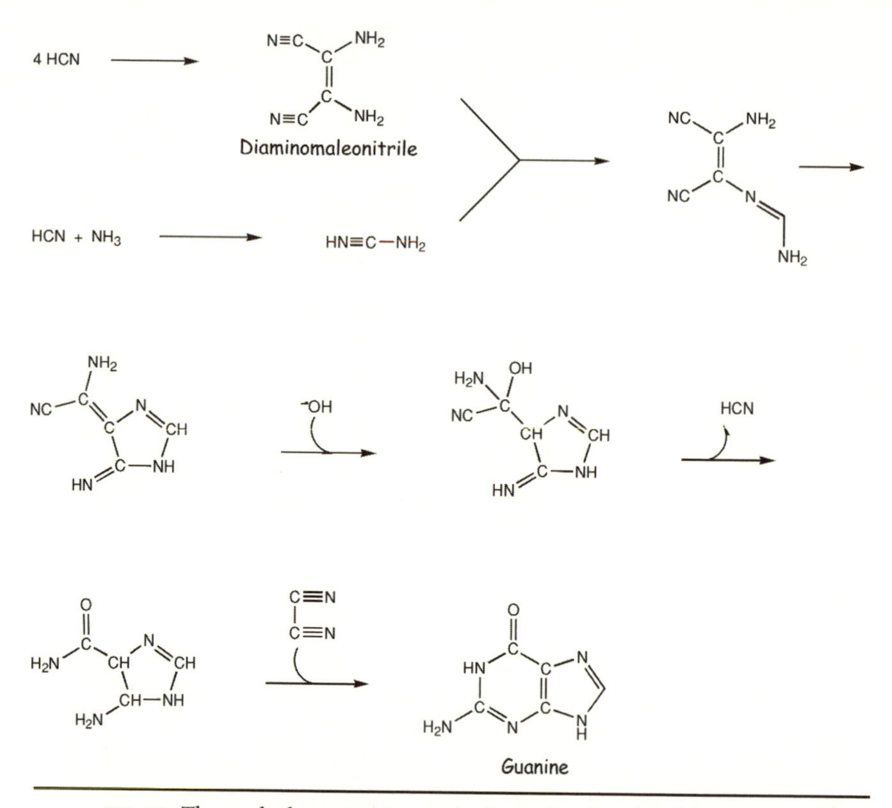

FIG. 4.8. The nucleobase guanine can be formed under plausible prebiotic conditions from ammonia and hydrogen cyanide. Even under optimized laboratory conditions, however, the yield of this reaction is quite low.

$C_n(H_2O)_n$. By definition, a sugar molecule contains three or more of these basic units. Sugars with five (pentoses; e.g., ribose) or six (hexoses; e.g., glucose) are particularly important in Terrestrial biology (fig. 4.9). In defiance of the historical name, however, the hydrogen and oxygen atoms in carbohydrates do not take the form of water molecules (i.e., sugars are not "hydrates" of carbon). Half of the hydrogens (plus one) are bound directly to a carbon, the other half (minus one) are bound to an oxygen atom to form an alcohol (an OH group), so sugars are also polyalcohols, or polyols. As the only exception to this rule, one of the carbons in every sugar is bound to an oxygen via a carbon-oxygen double bond to form an aldehyde or ketone functional group (the C=O group at the end or in the middle of the chain, respectively).

The trouble with sugars is that, even though Miller-Urey chemistry can produce simple alcohols, aldehydes, and ketones, it cannot produce

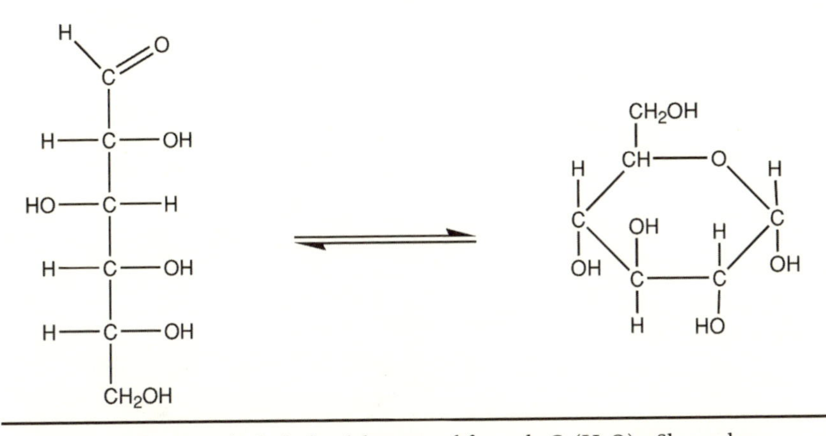

FIG. 4.9. Sugars are polyalcohols of the general formula $C_n(H_2O)_n$. Shown here is glucose, an aldohexose (*aldo* designating the double-bonded oxygen at the end of the chain; *hexose,* the six-carbon chain). Sugars such as glucose typically exist in two or more interconverting forms. On the left is the linear form, and on the right is a cyclized conformation produced when one of the alcohol groups attacks the aldol carbon. For glucose, the cyclic form dominates.

even the simplest, three-carbon sugars. Sugars are simply too large to be synthesized in the atmosphere by lightning strikes; long before the simplest sugar is produced, the precursor molecules would rain out of the sky. Sugar formation, then, must have occurred in the oceans or on land via the polymerization of the smaller precursors that Miller-Urey chemistry can produce. How might that have happened?

In 1861, the Russian chemist Alexander Butlerov (1828–86) described the formose reaction (fig. 4.10), which would later be hailed as the most plausible prebiotic route to sugars. The formose reaction starts with formaldehyde, which, with its formula $H_2C{=}O$, has the same carbon-plus-water structure as sugars; and although formaldehyde is not itself a sugar, sugars are, in a sense, formaldehyde polymers. Under the catalytic influence of bases, such as calcium hydroxide, formaldehyde dissolved in water polymerizes. The formose reaction starts with relatively concentrated solutions of formaldehyde (usually 1%–2%). After a short incubation with the base, the reaction accelerates rapidly, producing a peak yield of up to 50% sugars. On further incubation the sugar content decreases, however, because the sugars that are formed are relatively unstable in the presence of the strong base used to catalyze the reaction.

The formose reaction is thought to proceed via the two-carbon intermediate glycoaldehyde, which is formed from two molecules of the

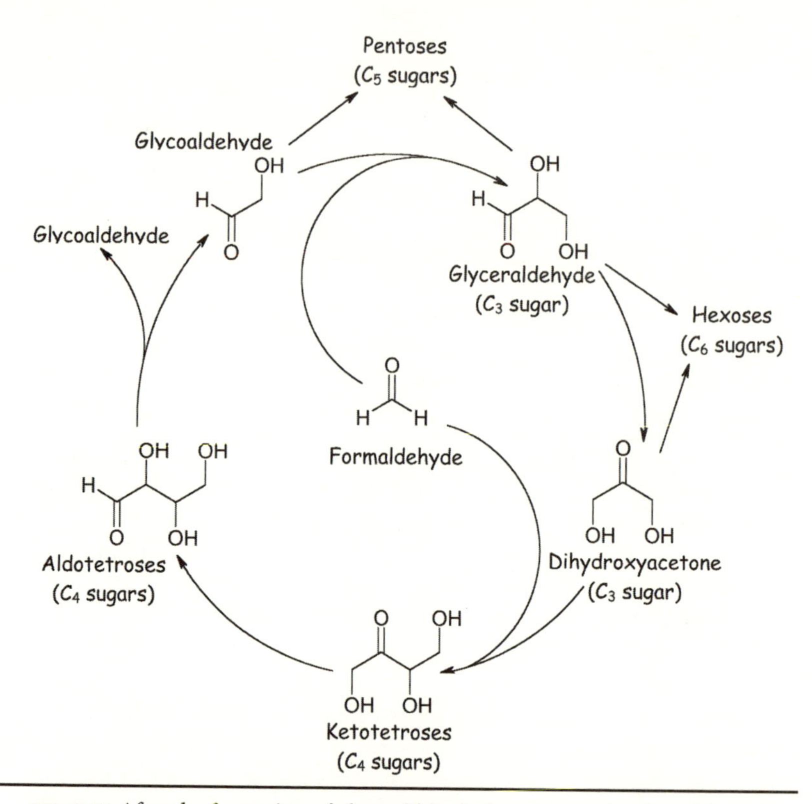

FIG. 4.10. After the formation of glyceraldehyde from two molecules of formaldehyde (not shown), the formose reaction synthesizes sugars from additional formaldehyde under reasonably plausible prebiotic conditions. The reaction is fairly nonspecific, however, and thus the yield of the important five-carbon sugar ribose (an aldopentose) is limited. The unlabeled vertices in the structural formulas represent carbon atoms.

one-carbon formaldehyde. Glycoaldehyde is also a catalyst in the formose reaction, and this autocatalysis accounts for the short delay observed before the reaction takes off; once a few molecules of glycoaldehyde have formed, they speed up the formation of yet more molecules of glycoaldehyde. Under basic conditions the glycoaldehyde in turn reacts with a third molecule of formaldehyde to form the three-carbon sugar glyceraldehyde. Glyceraldehyde, in turn, can undergo a base-catalyzed isomerization reaction that converts it into the other three-carbon sugar, dihydroxyacetone (these two sugars, which have their C=O at the end and in the middle of the molecule, respectively, are the only possible three-carbon sugars).

Glyceraldehyde and dihydroxyacetone are the two simplest sugars. From them, however, the formose reaction can produce many, much larger sugars. For example, two of these three-carbon sugars can combine via a well-known reaction called an "aldol condensation" (because it results in an *ald*ehyde-alcoh*ol*) to form most of the six-carbon sugars (hexoses). Alternatively, dihydroxyacetone can condense with another molecule of formaldehyde to form the four-carbon erythrulose, a ketotetrose. Erythrulose, in turn, isomerizes to form the aldotetroses (erythrose and threose), which can undergo a reverse version of the aldol condensation, splitting into two glycoaldehydes and starting the cycle anew. Of note, the isomerizations and aldol condensations thought to occur in the formose reaction mimic the chemistry by which current-day Earth life produces sugars from smaller precursors. Whether this coincidence represents a historical artifact of prebiotic chemistry that was later hijacked by life or simply represents the easiest chemistry by which sugars can be made (and thus the most likely chemistry for evolution to independently discover) remains an open question.

While the formose reaction has some of the traits we are looking for—it is spontaneous and can produce high yields—it is not without its problems as an explanation for the prebiotic synthesis of sugars. The first is that concentrated formaldehyde is difficult to come by. Formaldehyde can be synthesized by Miller-Urey chemistry (fig. 4.3). But given the reactivity of formaldehyde, it seems difficult for abiotic processes to generate enough formaldehyde to drive the reaction forward. Moreover, even if sufficient formaldehyde were available, the problem would remain that the formose reaction is extremely nonspecific and thus produces only small quantities of each of many different types of sugars. In particular, the formose reaction produces very little ribose, which in many regards is the most fundamental sugar for life on Earth and is intimately coupled to the most promising theories of life's origins. Worse still, what little ribose is produced by the formose reaction is relatively unstable. Even at modest temperatures, the half-life of ribose in water is just decades. It is thus hard to imagine that much ribose could build up simply via the formose-induced polymerization of formaldehyde in a primordial ocean.

Fortunately, at least a partial solution to the "ribose problem" has been demonstrated in the laboratory. The formose reaction is traditionally catalyzed by calcium hydroxide, which is a fairly powerful base, and very little ribose accumulates under strongly basic conditions because the sugars produced polymerize further into a complex, brown tar. Steve Benner at the University of Florida has shown, however, that borate, which occurs in nature as borax or colemanite, stabilizes ribose under the conditions in which the formose reaction can proceed. Borate, it turns out, forms a cyclic ester (a ring-shaped molecule with two oxygen-boron bonds) with two of ribose's hydroxyls, rendering the sugar unreactive and terminating the reaction. Of course, while this dramatically pushes up the yield of ribose in the formose reaction, it does nothing to stabilize ribose once the free sugar is released from the boron. The chemical half-life of ribose remains extremely short relative to the lengths of time we think prebiotic chemistry requires to build up usable quantities of life's precursors.

Lipids, long-chain molecules consisting almost solely of carbon and hydrogen, also play fundamental roles in biochemistry in, for example, the formation of the cell membrane. And yet, as for sugars, Miller-Urey-type reactions do not produce lipids. The reason is that Miller-Urey chemistry is generally oxidizing (remember: oxidation means removal of electrons, which is exactly what the radicals are doing; see fig. 4.3), and thus, while it readily produces relatively oxidized, oxygen- or nitrogen-containing compounds such as alcohols and amino acids, it does not produce the long, unoxidized carbon chains of lipids. The only way long carbon chains can form via Miller-Urey chemistry is by radical-radical reactions in which methyl radicals add to a growing carbon chain. As the chance of the radical reacting with a water or ammonia molecule, rather than with another methyl radical to grow the carbon chain, is reasonably high, large carbon compounds become rarer with increasing chain length. An additional reason Miller-Urey chemistry does not form long-chain carbon compounds is that they are insufficiently volatile; Miller-Urey chemistry is gas-phase chemistry and, as mentioned above, molecules with more than about three or four carbons fall out of the atmosphere as liquids or solids.

Perhaps the most feasible theory described to date for the prebiotic formation of lipids is that they were synthesized via the reduction of that other class of larger carbon-containing compounds: the sugars. But again, these compounds are not volatile and thus the reduction cannot occur in the atmosphere. So where can this happen? Deep in the planetary crust, where reduction can be catalyzed by the iron mineral troilite (FeS). In the presence of the reducing gas hydrogen sulfide (remember: the crust is reduced—at least in this model of prebiotic chemistry), troilite is a strong reducing agent. In fact, it is a strong enough reductant to produce hydrogen, to reduce alkenes (compounds containing carbon-carbon double bonds), alkynes (carbon-carbon triple bonds), and thiols (SH-containing molecules) to saturated hydrocarbons, and to reduce ketones to thiols (fig. 4.11). And even today, some 4.6 billion years after our planet formed, the Earth remains sufficiently geologically active that

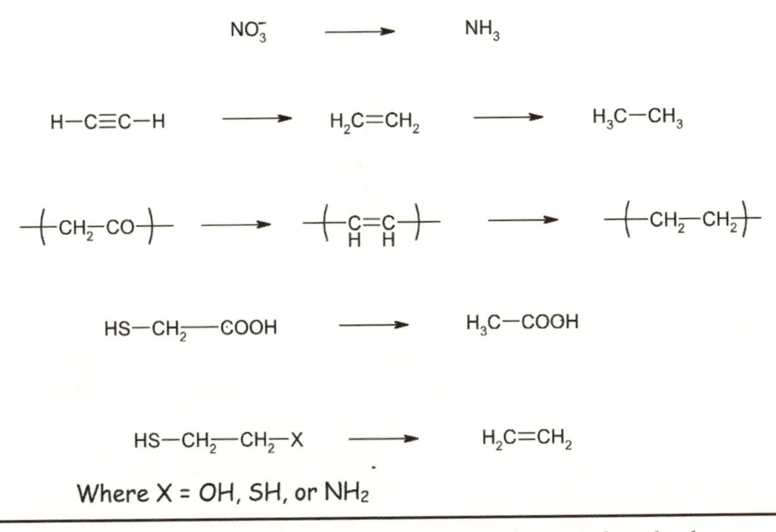

FIG. 4.11. A wide variety of reduction reactions can be carried out by the catalyst iron sulfide (FeS, which forms the mineral troilite) at elevated temperatures. These conditions might occur deep in a planet's primordial crust, as ocean water is filtered past highly reduced crustal rocks. Such chemistry might be the prebiotic source of lipids, which are critical for biology and are not produced via Miller-Urey chemistry.

deep sea vents in the mid-ocean ridges recycle an entire ocean's volume of water every 8 million years, and thus catalytic reduction deep within a planet's crust is a distinct possibility.

Potential Non–Miller-Urey Sources of Life's Building Blocks

Given the possibility that primordial atmospheres are too oxidized to promote Miller-Urey chemistry, we should note that a small but vocal group within the astrobiology community has argued that the prebiotic precursors to life are not formed in situ but delivered to terrestrial planets from space, where they can be synthesized under more reducing conditions.

Comets are known to contain ammonia, methane, and water (as solids, all of which are referred to as "ices" by planetary scientists), as do, perhaps, the asteroids that formed out beyond the "snow line." Both types of objects are also subjected to significant radiation (in the form of solar UV and cosmic rays) and therefore may contain Miller-Urey-type small molecules. Do they? Each of the half-dozen spacecraft that have visited comets to date has flown by at such high velocities that the delicate molecules we are interested in would not survive intact their encounter with the craft's instruments, and thus we do not yet know. This will change in 2014 when the European Space Agency's *Rosetta* mission, which is now en route, falls into orbit around the comet 67 P/Churyumov-Gerasimenko. And even if life's precursors do exist on comets and in the outer asteroid belt, a second question remains: could these delicate organic molecules survive impact with a planet?

The answer seems to be a resounding yes. Indeed, such delivery of organics has been observed in historical times in falls of carbonaceous chondrite meteorites. Carbonaceous chondrites are water- and organic-containing meteorites thought to arise in the outer half of the asteroid belt (beyond the "snow line" in which water condensed in the presolar nebula). Because they are made up of clays and other highly hydrated minerals, carbonaceous chondrites typically disintegrate soon after landing on our warm, wet planet, and thus historically they have been difficult to study. An important counterexample, however, arose in late 1969 when a large carbonaceous chondrite exploded in the air over Murchison, Australia. The fragments of the Murchison meteorite were quickly collected under dry conditions, allowing their leisurely study in the laboratory. Hot water extractions of the water-soluble organic material in the Murchison meteorite indicated that it contains a wide range of amino acids. In fact, some seventy different amino acids have been identified to date, including at least six of the twenty proteogenic amino acids found in life on Earth (table 4.4). Probably not coincidentally, many of the amino acids most readily formed in the Miller-Urey chemistries are found in abundance in Murchison material, suggesting perhaps that the Strecker synthesis was occurring in the presolar nebula. Radio astronomers have identified the unambiguous spectral signatures of a number of small organic compounds (the most complex being the two-carbon, sugar-like molecule glycoaldehyde) in nebulae thousands of light-years from Earth, so it is a good bet that prestellar nebulae contain such compounds.

Better still, the organic inventory of carbonaceous chondrites isn't just limited to amino acids. Dave Deamer, at the University of California, Santa Cruz, has extracted lipids from meteoritic material. When added to freshwater these lipids spontaneously form hollow spheres, called vesicles, reminiscent of cell membranes. On the other hand, Deamer's experiments have led him to question the supposition that life originated in the ocean. It turns out that, in saltwater, the lipids simply clump together without forming a hollow sphere, so Deamer has theorized that a freshwater pond would have been a more hospitable environment for life's origins.

So it seems that meteorites and, no doubt, comets can deliver bio-

TABLE 4.4
Amino acids detected in the Murchison meteorite

Isovaline	β-Aminoisobutyric acid
α-Aminoisobutyric acid	Pipecolic acid
Valine	**Glycine**
N-Methylalanine	β-Alanine
α-Amino-*n*-butyric acid	**Proline**
Alanine	γ-Aminobutyric acid
N-methylglycine	**Aspartic acid**
N-ethylglycine	**Glutamic acid**
Norvaline	

Note: Proteogenic amino acids in bold type.

logically relevant molecules from the outer reaches of a solar system to the rocky inner planets. But the question remains: could they do so in sufficient quantity to be relevant to the origins of life? Jeffrey Bada, a geochemist at Scripps Institution of Oceanography near San Diego, California, long argued that the delivery of organic building blocks from space is insufficient. He arrived at this conclusion by searching for the amino acid α-aminoisobutyric acid (AIB) in places like Antarctica and Greenland, where meteoritic material falling on snow and ice can build up over significant timeframes. AIB is rare on Earth because, unlike the proteogenic amino acids, it is not associated with life. But AIB is efficiently synthesized via Miller-Urey chemistry and is a major organic component of the Murchison meteorite. Bada found less than 0.05 milligram of AIB per square centimeter of ice surface, leading him to claim that this extraterrestrial molecule—and by extrapolation others—could not have built up to high enough concentrations to contribute much to the primordial soup.

More recently, however, Bada had to reevaluate this claim when he discovered that enormous quantities of buckyballs had arrived on Earth intact from outside the Solar System. Buckyballs, known more formally as fullerenes, are large, spherical molecules consisting of pure carbon (they adopt the appearance of a geodesic dome and thus were affectionately named after Buckminster Fuller, the engineer-inventor who popularized such structures). But buckyballs are a common component of soot, so how did Bada know that the buckyballs he found were of extraterrestrial origin? The spherical structure of a buckyball allows it to trap atoms inside. With his associate Luanne Becker, Bada explored a 2-billion-year-old impact site in Ontario, Canada, that contained about a million tons of carbon-rich buckyballs, many of which had entrapped helium atoms. And while helium is the second most common element in the Universe, it is rare on terrestrial planets (due to its light weight and chemical inertness, it tends to be lost to space over geological time). Thus it seems likely that these buckyballs, and presumably large amounts of other organic compounds, were indeed delivered from the outer Solar System by impacts. The growing consensus is now that both extraterrestrial delivery and in situ Miller-Urey chemistries contributed to the formation of the rich, prebiotic soup of organic materials necessary for life to form.

Prebiotic Polymerization

In our exploration of prebiotic chemistry we still face a significant hurdle. While we understand where many of the *monomers*—individual amino acids and nucleobases—come from, monomers do not equal life. Even nucleobases are relevant to life only in the context of three-part, covalent combinations of sugars, bases, and phosphate termed *nucleotides.* The reactions that synthesize nucleotides or nucleotide-like monomers from simpler precursors, and the reactions that polymerize such monomers into complex polymers, are both presumably critical steps in the origins of life, not only on Earth, but anywhere. But these reactions generally consume significant energy and thus are extremely unfavorable. How, then, might they have spontaneously occurred on a prebiotic planet? We don't really know, but several (admittedly rather speculative) mechanisms have been proposed.

Nucleobase-sugar combinations come in two basic forms. The nucleosides consist of a nucleobase covalently linked to a sugar, and nucleotides consist of nucleosides plus one or more phosphates (fig. 4.12). Nucleosides and nucleotides are synthesized from free sugars, nucleobases, and phosphates via *dehydration* reactions. That is, the chemistry that links these pieces together proceeds with the loss of water. Because of this, these reactions are unfavorable when they take place in liquid water, as the H_2O concentration of water is a whopping 55.5 mol/L—remember: water has the highest molar density of almost any substance. With all this water around in aqueous environments, reactions tend to want to go in the direction of hydration (which tends to break polymers apart) rather than dehydration, which is why the formation of nucleosides and nucleotides is unfavorable and must somehow be driven. So a key question is: how might this chemistry be encouraged under plausible prebiotic conditions?

A relatively straightforward route to the synthesis of the nucleoside inosine, which contains the nucleobase hypoxanthine, has been demon-

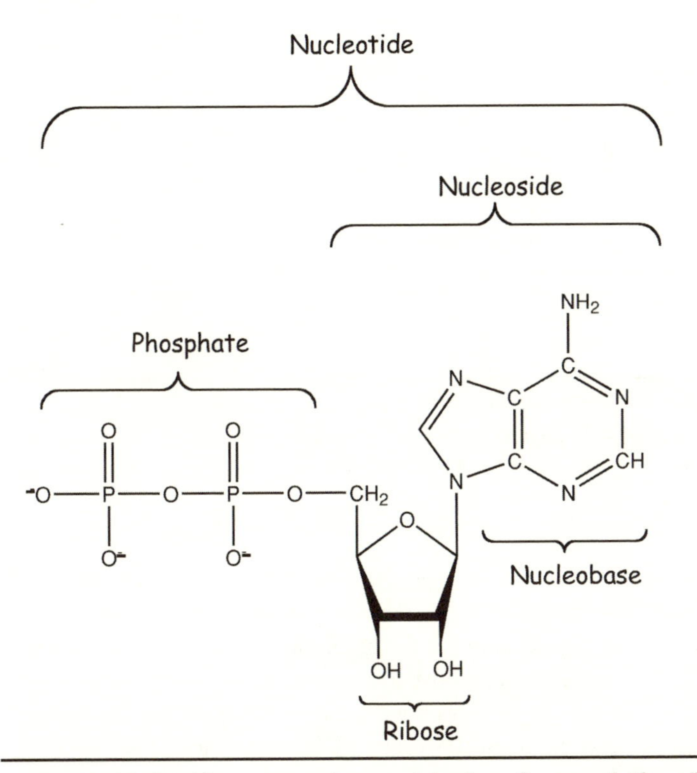

FIG. 4.12. Nucleosides are a covalent combination of a sugar (either ribose, as shown here, or deoxyribose, which we discuss in the next chapter) and a nucleobase. A nucleotide is a nucleoside plus one or more phosphate groups. Shown here is the nucleotide adenosine diphosphate (also known as ADP). The unlabeled vertices represent carbon atoms.

strated by Leslie Orgel, at the Scripps Research Institute. If a mixture of hypoxanthine, ribose, and magnesium ions is heated to dryness, inosine forms in good yield (fig. 4.13). Even better, the nucleoside is produced in the so-called β configuration, in which the nucleobase is on the opposite side of the sugar ring from the sugars of two other hydroxyl groups, which is the structure observed in biologically relevant nucleosides on Earth. How does this reaction work? The magnesium ion binds to two of the hydroxyls on the ribose, thus "activating" the carbon on which the nucleobase is to reside. A reactive nitrogen atom in the hypoxanthine attacks the activated carbon (from the face opposite the magnesium, thus accounting for the formation of the β structure), forming a nucleoside. Heating to dryness drives the linkage chemistry forward by driving off the water that is liberated as the reaction progresses. Unfortunately—

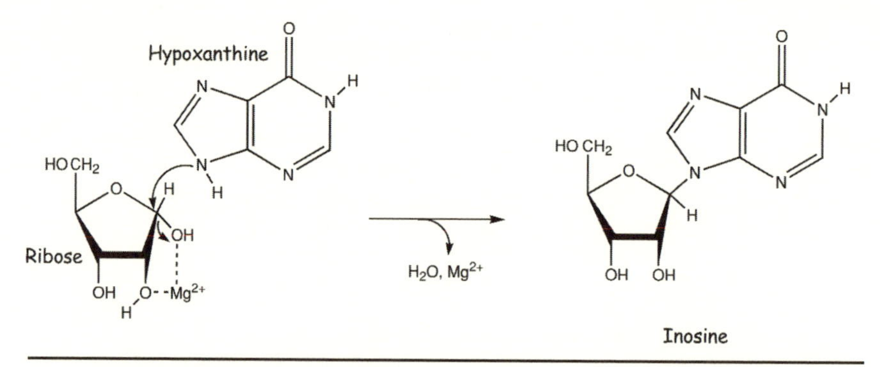

FIG. 4.13. Orgel has shown that, with magnesium ions as a catalyst, the nucleoside inosine forms spontaneously from ribose and the nucleobase hypoxanthine. No other nucleobase, however, has been shown to undergo this reaction.

isn't it starting to seem as though there's always an "unfortunately" in this business?—this reaction is neither general enough nor specific enough to plausibly account for the prebiotic synthesis of nucleosides. It is insufficiently general because hypoxanthine is the only nucleobase that is reactive enough to form nucleosides by this mechanism, and of the four nucleobases for which we have plausible prebiotic syntheses, it's the one that is least relevant, at least for contemporary Earth life. Orgel's mechanism is also insufficiently specific: many of the range of possible five-carbon sugars react equally well, which would waste precious nucleobases without producing the desired ribonucleoside. What we really need is another means of activating the sugar for attack that is specific for ribose and generalizable to all of the nucleobases. To date, however, no such chemistry has been described.

But free nucleosides are not the stuff upon which life is founded. Instead, life is built around polymers of nucleotides, long chains of nucleobases bound (in the case of RNA) to ribose molecules that are in turn linked, hydroxyl group to hydroxyl group, via intervening phosphates. These polymeric nucleic acids are higher in energy than monomers because the bonds that link the monomers together in these compounds are themselves unstable; they too are the products of dehydration reactions. To date, a number of potential prebiotic chemistries have been proposed by which a nucleoside could be converted into a high-energy nucleotide that, in turn, could be polymerized to form chains. Perhaps the most compelling of these is the use of high-energy polyphosphates, which is in fact the system Terrestrial biology uses today to store and shuttle energy.

The simplest nucleotides are the monophosphates, in which a single phosphate is linked to a hydroxyl group on the nucleoside's ribose. These can be synthesized in up to 60% yield by nothing more than simple heating of a solution of nucleosides, phosphoric acid, and urea (which acts as a catalyst). Unfortunately, however, nucleoside monophosphates are not the sort of high-energy compound that can be polymerized to form polynucleotides. In contrast, a string of phosphates linked via oxygen bridges is higher in energy because, at neutral pH, each phosphate takes on a negative charge, and the negative charges on adjacent phosphates repel one another. Contemporary Earth life utilizes the energy available from this repulsion to drive the polymerization of RNA and DNA (not to mention countless other biochemical reactions within our cells). And it seems it just might be possible for nucleoside polyphosphates to be synthesized under reasonably plausible prebiotic conditions.

The postulated reaction involves the triphosphate molecule trimetaphosphate, which consists of three molecules of phosphoric acid that have become linked together by, perhaps not surprisingly, dehydration reactions. Heating a dry mixture of nucleoside monophosphates with trimetaphosphate in the presence of the catalyst magnesium (again, the heat drives off water, propelling the reaction forward) can produce nucleoside polyphosphates in good yield. When water is added to these compounds, they rapidly hydrolyze to form nucleoside triphosphates (which hydrolyze only slowly to form di- and monophosphates). Unfortunately, however, the relevant phosphate compounds are quite rare on Earth, which is why phosphate is usually the limiting mineral in freshwater ecosystems and why, in turn, phosphate detergents were wreaking havoc on rivers and lakes until they were banned in the 1970s. Whether phosphates are similarly rare on primordial terrestrial planets is not known.

Of course, just because polyphosphate plays the role of activator on the contemporary Earth doesn't mean that it's the only way to activate nucleosides or, indeed, that it was the activation chemistry employed in the origins of life. For example, some twenty years ago Orgel proposed that the small, nitrogen-containing ring compound imidazole could react with phosphate to form a phosphorimidazolide, a high-energy nitrogen-phosphorus adduct that activates the phosphate for further reactions. Unfortunately, however, while phosphorimidazolides can be synthesized in the laboratory, no plausible prebiotic route to their synthesis has been described. And if some alternative activation chemistries were in play during the origins of life, they have not yet been identified.

Activated nucleotides are the starting materials from which RNA polymers are synthesized. How such polymerization might take place

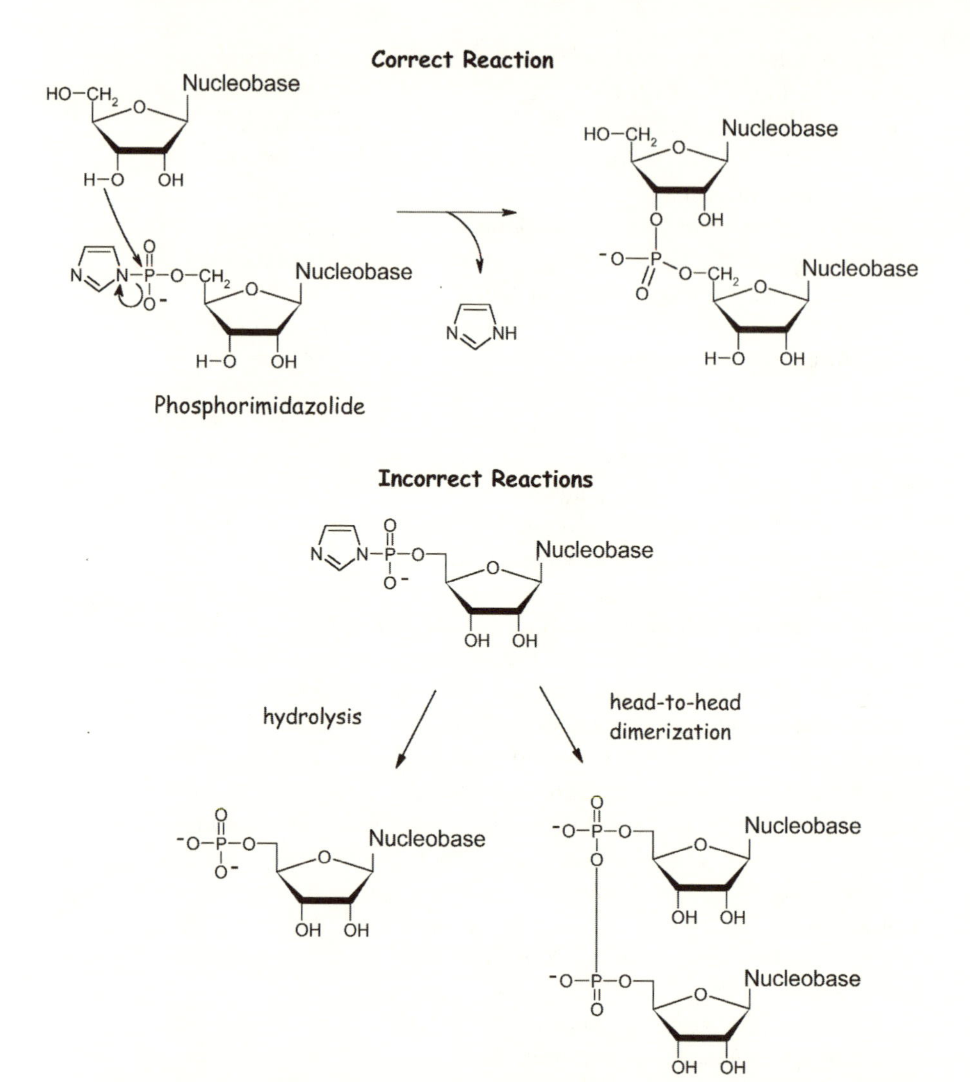

FIG. 4.14. Activated nucleotides, such as the phosphorimidazolide shown here, can be polymerized in solution. The yield of correctly linked RNA polymers is low, however, due to the multitude of possible side reactions.

has been the subject of extensive study. One possibility is simply the spontaneous polymerization of phosphorimidazolides (or other, still unknown, activated nucleotides) in solution. This reaction, however, tends to be extremely inefficient. The problem is that each nucleotide (ribonucleotide, that is) has two free hydroxyl groups where the linkages can be formed (fig. 4.14), and only one of them forms the correct structure employed in life (or, at least, life here on Earth). In fact, for the activated base adenosine phosphorimidazolide, the incorrect linkage is six times more likely to form than the correct one. The problem of incorrect linkages, however, may be surmountable by the introduction of mineral catalysts. For example, James Ferris, at Rensselaer Polytechnic in upstate New York, has shown that RNA polymers up to fifty-five monomers long can be synthesized on the common clay montmorillonite and that the product contains mostly the correct linkages; it seems that the surface of the clay binds the growing polymer and carefully directs the addition of each new monomer.

Alternatively, the polymerization can be *template directed.* RNA (like its relative DNA) can bind to a complementary sequence to form a double helix. This suggests that an existing RNA polymer could act as a template to direct the specific polymerization of a new strand of RNA. Using polycytidine as a template to direct the polymerization of activated guanosine, Orgel found that the correct linkage is favored 2:1 in the final polymerized product. The presence of zinc increases the yield of polymerized product without harming the yield of proper linkages. In fact, when the activator 2-methylimidazole is employed instead, and in the presence of magnesium, sodium, and zinc ions, a template containing both uracil and cytosine can be copied quite efficiently as long as cytosine dominates the template's composition. If the cytosine and uracil contents approach equality, the yield of polymerized material plummets. This limitation presents a serious obstacle to the formation of a self-replicating system from the simple, template-directed polymerization of RNA; as we discuss again in the next chapter, a cytosine-rich sequence that serves as an efficient template will produce cytosine-poor products that cannot, in turn, serve as new templates.

So, it seems the synthesis of RNA polymers might be possible under plausible prebiotic conditions, but is there chemistry by which activated nucleotides can be synthesized abiologically from nucleobases and ribose? Sadly, the answer is still no. Indeed, the situation may be even worse than we've let on. Not only is the synthesis of nucleotide polymers an (energetically) up-hill battle (see sidebar 4.2), but staying "up hill" is equally difficult. For example, the half-life of the phosphodiester linkages in RNA is less than 1,000 years in water at the freezing point, and less than 1 day at 35°C. And given that even a modest-length RNA polymer contains hundreds of phosphodiester bonds, the lifetime of an intact polymer is much, much less than the lifetime of any one linkage. This observation has led to the speculation that more-stable, RNA-like polymers were instead involved in the origins of life. We return to this important issue in the next chapter.

SIDEBAR 4.2

Weighing the Probabilities

How likely is it that a thick, rich soup of amino acids and activated nucleotides can form on a primordial, terrestrial planet? If we limit ourselves to currently well-understood chemistry the answer is, unfortunately, "not very." For example, no one has yet postulated a chemically plausible scenario by which most of the nucleosides can be formed from nucleobases and ribose, and the serious mismatch between the rates at which cytosine and ribose might be produced and the rates at which they are destroyed (and thus, in net, their equilibrium concentrations) suggests that these equally critical compounds would be quite rare.

The problem of the stability of likely biopolymers is similarly acute. The polymerization of RNA, for example, is an energy-consuming process, and thus the formation of longer and longer RNA polymers becomes less and less likely unless the polymerization reaction can be coupled with some other energy-liberating reaction. This is why searches for chemistry that might "activate" nucleosides have been so intense; without such activation, it is difficult to see how the relevant reactions could proceed to generate even trace amounts of polymers of sufficient length.

Of course, even up-hill chemical reactions can proceed—they just do so at lower and lower yield as the energy consumed in the reaction increases. Boltzmann's law states that the ratio of products to starting material for the conversion of molecule A into the product molecule B is proportional to $e^{-\Delta G/\mathrm{k}T}$, where ΔG is the energy difference between molecules A and B, and $\mathrm{k}T$ is the mean energy of molecular collisions at a given temperature T. Because the fraction of molecules undergoing a given chemical transformation varies exponentially with the energy consumed in the reaction, the polymerization of unactivated nucleotides to form even short polymers is astronomically unlikely. But it is not impossible. And thus the origins of life may once again be tied up in the anthropic principle (i.e., we do not know how probable the formation of life was, only that the probability was not zero). Alternatively, of course, the solution to this conundrum may lie in our incomplete knowledge. Specifically, while it is fairly advanced, our knowledge of chemistry in general and of the chemistry of primordial, rocky planets in particular is hardly exhaustive. Thus there may well be high-yield synthetic routes by which all of the components of the putative primordial soup could have formed. The jury, in short, is still out.

Conclusions

We've come a long way in fifty years, but perhaps not as far as Miller would have thought while marveling over his results back in 1953. Thanks to his work and that of those who followed, we understand in detail how many of the key molecular elements of life as we know it might have been created in the prebiotic world. But just as many questions about prebiotic chemistry remain. Will they be answered in the next fifty years? Let us hope so; and let us be assured that, if they are not, it won't be for lack of trying!

In the early 1960s Harold Urey went on to head the Chemistry Department at the newly founded University of California, San Diego, and he brought along his protégé Stanley Miller. Miller spent the next forty years pursuing origins-of-life research, publishing several hundred papers and numerous books on the topic, and spawning a whole community of scientists pursuing the same goal. And yet, now fifty years after Miller's original experiment produced a fascinating framework with which to think about prebiotic chemistry, many key details of the process remain to be worked out.

Further Reading

Current thoughts on prebiotic chemistry. Orgel, Leslie E. "Prebiotic chemistry and the origin of the RNA world." *Critical Reviews in Biochemistry and Molecular Biology* 39 (2004): 99–123.

A textbook of prebiotic chemistry. Zubay, G. *Origins of Life on the Earth and in the Cosmos.* San Francisco: Academic Press, 2000.

CHAPTER 5

The Spark of Life

Throughout most of history (and, we presume, prehistory), it was assumed that life could arise spontaneously from inanimate matter. After all, common sense dictates that, for example, maggots arise spontaneously in spoiled meat, and throughout the Middle Ages it was widely held that old rags mixed with wheat could give rise to fully formed adult mice.

By the early nineteenth century, however, the theory of spontaneous generation was in doubt. The mouse-rag idea, for instance, had by then largely been discounted, as it was clear that macroscopic organisms like ourselves arise only as the offspring of parents of the same species. But what of the smaller, simpler organisms that cause fermentation and spoiling? For example, do the organisms that make milk go sour arise spontaneously in the milk?

Interest in this question grew over the next half-century until, in 1859, the French Academy of Sciences founded a prize of twenty-five hundred francs for the scientist who could lay the matter to rest by conclusively proving or disproving spontaneous generation. The prize was quickly won by a relatively young chemist recently appointed to a professorship in Paris, by the name of Louis Pasteur (1822–95). His decisive evidence, published in 1861, was provided by his now famous experiments with curve-necked flasks. Pasteur placed various liquids—sugar solutions, urine, and beet juice—in swan-shaped flasks whose long, curved necks allowed communion with the air but would prevent, he postulated, small living particles (our bacteria) from dropping into the liquids with the settling dust. In the flasks that he sterilized by boiling, Pasteur found that nothing happened. In contrast, the liquid in the unboiled flasks rapidly spoiled. Of course, that could simply have meant that boiling had somehow made the liquid unpalatable for the little organisms. A decisive result was provided when Pasteur snapped off the swan-shaped necks, allowing dust to settle into the previously boiled organic brews. Not surprisingly, the liquid then spoiled, demonstrating that its prior sterility reflected the absence of spontaneous generation, rather than ruined media that could no longer support life, and that the organisms responsible for the spoiling had settled out of the air. Pasteur proclaimed to the French Academy: "Never will the doctrine of spontaneous generation recover from the mortal blow of this simple experiment" (quoted by Fry in *The Emergence of Life on Earth,* 2000).

Pasteur's conclusions, though, were in conflict with a theory published just three years earlier by a scientist on the other side of the English Channel: Darwin's theory of the origins of species. Charles Darwin (1809–82) had obviated the need for God in the creation of species by showing how one species could slowly transmute into another. But Darwin's theory left open the question of how the first species arose. At the very least, the *first* species cannot have arisen via the transmutation of some earlier species and thus cannot have arisen without some form of spontaneous generation. Darwin was rather coy on the subject, both noting (as quoted by Davies in *The Fifth Miracle,* 1999) that "I have met with no evidence that seems in the least trustworthy, in favor of so-called Spontaneous Generation," and speculating that life may first have arisen "in some warm little pond, with all sorts of ammonia and phosphoric salts, light, heat, electricity *etc.*" Resolution of this paradox was, and remains, one of science's greatest mysteries.

Panspermia

Within a decade of Pasteur's publication, several prominent European scientists—most notably the English physicist Lord Kelvin, of absolute temperature fame, and the Prussian physicist Hermann von Helmholtz (1821–94), of, for example, the Gibbs-Helmholtz equation in chemical thermodynamics—suggested a possible work-around. Can we not avoid the difficulty of spontaneous generation if life had originated elsewhere and been transported to the Earth through space? Their early speculations on this, the "panspermia hypothesis," were fleshed out in great detail in a widely discussed body of work by the Swedish chemist and 1903 Nobel laureate (again—the prize gives you a lot of leeway to pursue wacky ideas) Svante Arrhenius (1859–1927), who was already famous for relating the temperature-dependence of chemical reaction rates. Originally in a 1903 journal article and then expanded upon in a popular book some five years later, Arrhenius argued forcefully that life, in the form of hardy, dormant spores, could survive in the cold dark vacuum of space for long enough to be transferred between the stars.

In Arrhenius's version of events, bacterial spores can escape from the upper atmosphere of an inhabited planet and then be launched into interstellar space by the pressure of light; photons have momentum and thus can accelerate something as small as, for example, a bacterial spore. Eventually, some of the spores fall upon another planet (Arrhenius estimated that sunlight could push a bacterial spore from the Earth to Mars in twenty days, and from the Sun to the nearest star in as little as nine thousand years), where they inoculate the virgin world. In the century since he first published his work, others have explored Arrhenius's hypothesis in much more quantitative detail. The well-known astronomer Carl Sagan (1934–96), for example, calculated that a space-borne spore must be less than 0.5 micrometer (half a millionth of a meter) in diameter before its surface-to-mass ratio is large enough that light can accelerate it outward against solar gravity and propel it out of the Solar System and noted that only the very smallest of Earth's bacteria, however, are this small. Sagan also described the various types of stars that spores could be blown from and to (from bright stars, which endow the spores with significant speed, to the planets of dimmer stars). He could not, however, come up with a solution to the vexing problem of radiation. Even the heartiest of Earthly bacteria would be destroyed within a day of leaving the Earth's protective atmosphere by the Sun's harsh ultraviolet light. And even if one postulates a super-sunburn-resistant bug, cosmic radiation would kill even this traveler long before it could leave the Solar System.

Putting the problem of being fried in transit aside, there is a more fundamental problem with the panspermia hypothesis: while it might explain the origins of life on Earth, it does not answer the more fundamental question of how life arose from inanimate matter in the first place. And thus, even if life could survive a trip between the stars, we astrobiologists can't sweep the mystery of life's origins under the rug simply by saying it took place elsewhere. Arrhenius himself sidestepped the issue of how life arose in the first place by suggesting that it might be eternal; after all, at the time it was assumed that the Universe was immortal, so why couldn't life have always existed as well? This argument was raised again half a century later by Fred Hoyle, the disparager of the Big Bang and the champion of his alternative cosmology of "continuous creation," which postulated that the Universe had no beginning (or end). But alas, continuous creation has since been thoroughly debunked. It is now well established that the Universe is "only" 13.7 billion years old. Somewhere, sometime during the last 13.7 billion years life must have arisen spontaneously; if not here, then somewhere else. And so the question remains: how did it do so?

Theories of the Origins of Life

The detailed, scientific consideration of the origins of life began with the aforementioned Soviet and Scottish scientists, Aleksandr Oparin and J. B. S. Haldane, the latter of whom probably coined the phrase "primordial soup." Oparin proposed a "cells-first" origin of life; impressed by the cell-like appearance that oily organic materials can adopt in water, Oparin proposed that the physical structure of the cell came first, in the form of a suspension of oily droplets and hollow, water-filled oily "vesicles," together called coacervates. Oparin noted that, in particular, these water-filled vesicles could serve as a vessel in which life's chemistry could arise, isolated from potentially disruptive influences. Moreover, Oparin noted, under some conditions these vesicles can grow *and then divide* in a manner reminiscent of cell division.

But, although it does not always seem so, life is ever so much more than just swelling, water-filled vesicles of lipids; life as we have defined it requires genes and metabolisms. The dispute over which of the two came first has divided researchers in the field, fueling a debate that many observers have likened to the classic chicken-or-egg argument, with both the metabolism-first and genes-first camps simultaneously claiming they have the upper hand.

Metabolism First

The metabolism-first camp argues that the first life was formed from a primitive network of self-sustaining chemical reactions of monomeric organic molecules, catalyzed by either organic or inorganic catalysts. As this interlocked network of reactions "evolved" in complexity, genetic molecules were somehow incorporated and "metabolic life" developed into life as we've grown to love it. Seems plausible so far. But two serious—and as yet unanswered—questions lie at the heart of the metabolism-first theories. The first is: precisely what sort of self-sustaining chemical reaction networks? And the second: how did such a system acquire the genetic material that characterizes all life on Earth (and is believed likely to characterize any life anywhere) and that is intimately related to that oh-so critical characteristic of life, *evolution*?

Günter Wächtershäuser, a Munich patent lawyer who dabbles in origins-of-life chemistry in his spare time,* proposed just such a metabolic network in the late 1980s. Wächtershäuser postulates that an assembly-line network of chemical reactions, which he lovingly describes (as quoted by Orgel, 2000) as "two-dimensional chemi-autotrophic surface metabolism in an iron-sulfur world," was set up on the surface of catalytic minerals, such as iron sulfide, in hydrothermal vents deep beneath the sea. The hydrothermal vent environment drives the reactions

*Although he's not made as much of an impact as another German patent clerk who published the theory of relativity in his spare time, Wächtershäuser has nevertheless caused quite a commotion with his speculations.

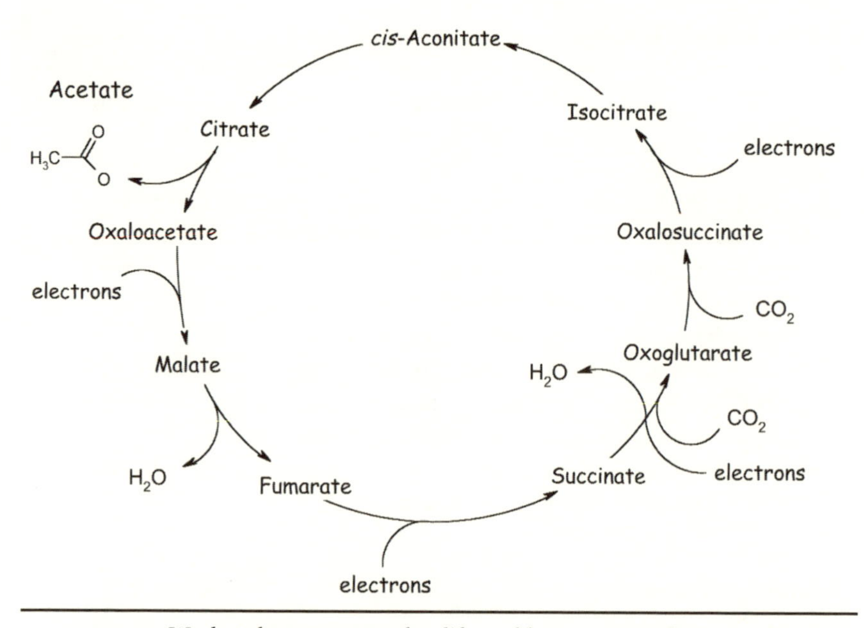

FIG. 5.1. Wächtershäuser argues that life could start via purely metabolic networks spontaneously arising on the surface of iron sulfide minerals. For example, he postulates that the reverse (reductive) Krebs (or citric acid) cycle can be catalyzed on FeS, which in net takes the reducing potential of hydrogen sulfide (electrons) and uses it to reduce carbon dioxide (entering at mid-right) into larger organic molecules such as acetate.

using the chemical disequilibria set up when hot, briny water emerges into the colder ocean.

Wächtershäuser proposes that the key set of reactions in his "metabolic life" formed a backward, reducing version of the Krebs cycle (also known as the citric acid cycle)—the central oxidative biochemical pathway in aerobic organisms (we explore this reaction in more detail in later chapters). He postulates that this, and a good deal of other, highly organized chemistry, can occur on the surface of iron sulfide minerals in the presence of hydrogen sulfide (fig. 5.1). Hydrogen sulfide (H_2S), a reducing gas produced in the crust, can convert the more oxidized iron sulfide mineral pyrite (FeS_2) into the more reduced mineral troilite (FeS). The troilite, in turn, can then catalytically reduce (i.e., provide electrons to) various organic molecules, converting itself back into pyrite. Wächtershäuser also suggests that the surface of iron sulfide would constrain distribution and orientation of the products of each reduction in such a way as to support a complex, self-sustaining sequence of metabolic reactions. Thus, speculates Wächtershäuser, the first life consisted of a metabolic network spontaneously formed on the surface of iron-sulfur minerals that used the reducing power in H_2S to reduce carbon dioxide to carbon-containing metabolites.

In support of his hypothesis, Wächtershäuser has shown that potentially biologically relevant reduction reactions can take place on iron sulfides. For example, in collaboration with Claudia Huber, at the Technical University of Munich, he has shown that, in the presence of iron and nickel sulfides, H_2S can efficiently reduce carbon monoxide to acetic acid—carbon *monoxide,* though, not the carbon *dioxide* that is essential to his postulated reductive citric acid cycle. The two researchers have similarly shown that, in the presence of ammonia and H_2S, iron and nickel sulfides can reduce a specific class of ketones, called α-ketoacids, into α-amino acids. So it seems that biochemistry-like reductive reactions can take place in the presence of mineral sulfides. It remains very much an open question, though, whether a complete metabolic network could spontaneously self-organize (and operate autonomously) on an iron sulfide surface under plausible prebiotic conditions. What, after all, would drive a collection of seemingly dissimilar chemical reactions to spontaneously self-organize?

Wächtershäuser has argued that the self-organization stems from the constraint of being adsorbed (or synthesized) onto the surface of a mineral. The organization imposed by the organized mineral surface would foster the formation of the network and increase its specificity. To date, however, no such mineral-guidance effect has been observed for any set of chemical reactions even remotely approaching the complexity that would be required to form a self-sustaining network that fed on simple precursors and synthesized larger molecules. Perhaps this is to be expected: any one mineral is unlikely to specifically catalyze more than one or two of the many distinctly different chemical reactions required for a metabolic network as complex as the Krebs cycle, much less catalyze *only* those reactions that are productive, and not competing side reactions that would disrupt the network. Moreover, even if such multicatalytic minerals did exist, there is no laboratory evidence that complex catalytic networks could spontaneously self-organize on their surfaces. Thus, while the iron-sulfur theory has stimulated discussion and research, it is not yet based on firm laboratory verification, or even laboratory-based hints that something might be possible under some, as yet to be defined, set of conditions.

Of course, we might still feel confident about the possibility of iron-sulfur life even in the absence of laboratory examples, if we could come up with an argument that an iron-sulfide-based chemical network could have evolved into us. That is, even though our interest is in defining the range of possible forms that life can adopt anywhere, not just on Earth, if we could argue that we had evolved from iron-sulfur life we'd feel more confident about the possibility of iron-sulfur-based metabolism as a precursor to life both on the early Earth and elsewhere in the cosmos. And while there are small clusters of FeS in some of the proteins that catalyze oxidative biochemistry, to date no clear route from there to here has been described. For example, why would nucleic acids, so central to our biology, provide a selective advantage for these systems? For that matter, since the FeS itself is not replicating, what if anything would provide a selective advantage—which is first and foremost about improving the ability to reproduce—for such a system? In the absence of compelling answers to these questions, it does not seem likely that our first ancestor was a gene-free metabolic network quietly chugging away on the surface of a piece of pyrite. And in the absence of a compelling demonstration of such chemistry in the laboratory, we must question whether any life anywhere could arise from such a start.

An alternative twist on the metabolism-first idea is that the first living organisms were not made up of the sort of complex, multistep networks that we typically envision when we say "metabolism," but instead were built from metabolic-like pathways involving lipid aggregates. Lipids are "water-hating" molecules (termed "hydrophobic" by the chemists), and thus, in solution, lipids tend to organize spontaneously into compact spheres, termed micelles, or into hollow, water-filled, membrane-bound balls called vesicles (soap is a lipid that spontaneously forms micelles around grease, solubilizing it and allowing it to wash down the drain). Even today, most biological membranes are not created from scratch, but rather by a growth-and-division process that is at least somewhat analogous to life. And the coacervates Oparin studied in the 1920s could sometimes be coaxed to adsorb smaller precursors and even divide. But replication alone does not life make. Is there a plausible chemical scenario in which a blob of membranes could be said to evolve? Doron Lancet, of the Weizmann Institute in Israel, and Dave Deamer, at the University of California, Santa Cruz, have argued that there might be (fig. 5.2).

The "lipid world" argument starts with the observation that lipids are an extremely diverse set of compounds. By way of example, modern eukaryotic cells contain three broad classes of lipids in their cell membranes: phospholipids, sphingolipids, and sterols. The former two differ in terms of the polar, water-loving "head group" attached to the water-hating fatty acid tail of the lipid. Sterols, however, represent a different

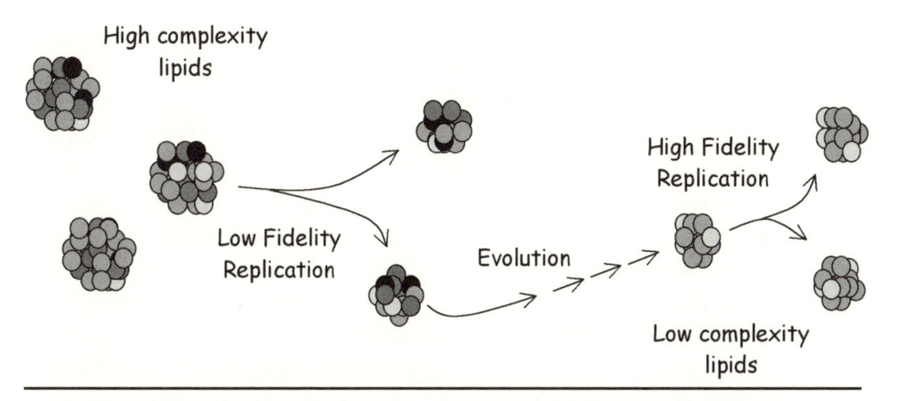

FIG. 5.2. The lipid-world hypothesis is predicated on the thought that there are so many different types of lipids that some (presumably very rare) lipid aggregates might exhibit the property of only adsorbing new lipid components in approximately the same ratio as the aggregates' existing lipids. Thus the composition would, more or less, "breed true" as an aggregate swelled and eventually split. If the breeding were not perfectly "true," and the mistakes—"mutations"—improved the fidelity or rate of the aggregate's growth and division, the system would evolve.

design, the most common example of which is cholesterol. Each of these three groups, in turn, can be subdivided by chemical details, such as the chemistry of the fatty acid tail, chemistry of the head group, oxidation state of the sterol rings, and so on. There are dozens of distinct types of lipids in the membranes of your cells, dozens and dozens of others in the membranes of prokaryotes, and countless more can be synthesized in the laboratory and may have been synthesized prebiotically. With such tremendous diversity, who can say what range of chemistries is possible? More specifically, notes Lancet, is it not possible that, among the myriad of possible lipid compositions in a simple vesicle, there might be some particular mixture that has the property of encouraging the adsorption of other lipids in exactly the right ratios to produce growth and division to form offspring similar to the original vesicle (i.e., aggregates that "breed true" and thus accurately pass down their traits to the generations that follow)? And if this is so, is it much more of a stretch to postulate that these lipid aggregates could sometimes accidentally take on some new lipid component that better suited them for their environments *and* also continue to breed true? Were such lipid aggregates possible, they could be said to be both reproducing and evolving. They could be said to be living things.

The lipid-world hypothesis nicely explains how a presumably very dilute primordial soup could spontaneously organize into complex structures; in order to minimize the extent to which their hydrophobic elements are exposed to water, lipids readily self-assemble into micelles or vesicles even when at extremely low concentrations. That said, the lipid-world hypothesis is still just that, a hypothesis. For example, to date, the micelle and vesicle chemistry we have observed in the laboratory just isn't very selective; no one has yet come up with a mixture of lipids that even comes close to breeding true, much less one that starts there and slowly evolves. The problem is that the aggregation of lipids to form micelles and vesicles is generally nonspecific; only relatively poor differentiation is observed. And thus there isn't much in the way of laboratory evidence supporting the ability of lipid vesicles to breed true, a key element of the lipid-world hypothesis. A second problem is shared with the iron-sulfur hypothesis: how would this evolve into us? While it is true that our cells are enclosed within membranes (and, indeed, membranes may have played an absolutely critical role in the evolution of life—just not a solo role!), the lipid-world chemistry is hardly reminiscent of our most fundamental metabolic and genetic chemistry. To quote its proponents: "A complex chain of evolutionary events, yet to be deciphered, could then have led to the common ancestors of today's free-living cells, and to the appearance of DNA, RNA and protein enzymes" (Segre and colleagues, 2001). Given that lipid-world-like chemistry hasn't yet been seen in the laboratory, and no one has come up with a plausible scenario in which a self-replicating lipid-only vesicle could have evolved into us, there doesn't yet seem to be any direct evidence suggesting that lipid life might be quietly inhabiting some distant corner of our galaxy.

Metabolism-first theories, such as the iron-sulfur world or the network of interactions inherent in the lipid-world model, face yet another, perhaps even more fundamental difficulty: gene-free networks are generally resistant to evolutionary change, because such change would require that multiple mutations occur simultaneously. Let's look at a network in which product A catalyzes the formation of product B, and vice versa (fig. 5.3). It's always possible that there is some "mutant" version of A, say A′, that is a *better* catalyst of the formation of B. Thus the mutant A′ would provide a selective advantage for the putative metabolism-only organism. But while A′ is a better catalyst, it is still likely to catalyze the formation of B, and B catalyzes the formation of A, not A′! And thus to truly provide an "inheritable" selective advantage, the mutant A′ must catalyze the formation of mutant B′ that, in turn, catalyzes the formation of A′. The evolution of networks fundamentally requires *multiple,*

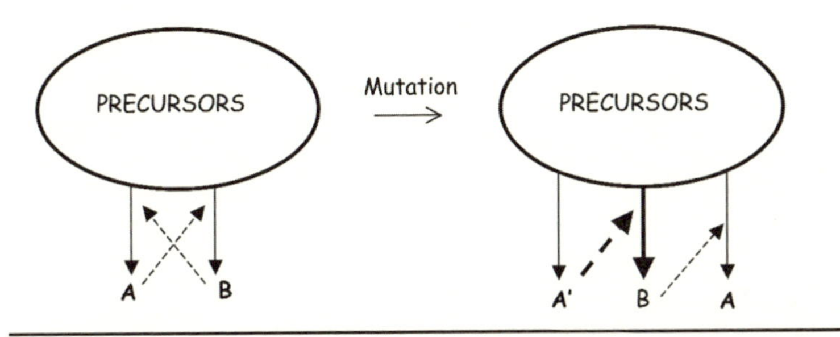

FIG. 5.3. A generic problem with metabolism-first theories of the origins of life is that gene-free networks are generally resistant to change. The problem is that multiple, simultaneous changes must occur, due to the interlinked nature of networks. If only one change occurs, the system will either stop or, as shown, revert to its original chemistry.

*simultaneous mutations.** And mutation probability does not change algebraically with the number of mutations, but geometrically: the formation of two simultaneous mutations is not *twice* as improbable as the formation of one, but instead is the *square of the improbability* of forming one mutation. And given that even single mutations are relatively rare—remember: if mutations occur too readily, few of the offspring will be viable—if we square the probability (or cube it, if it is a three-part network, and so on), the ease with which mutations produce a selective advantage diminishes very rapidly. Thus it is extremely difficult for complex networks to evolve a slow accumulation of stepwise mutations under the influence of selective pressures; at a fundamental level, it seems unlikely that networks could give rise to life before first acquiring genes, because only genes *allow for the stepwise formation of inheritable mutations!*

Genes First

The genes-first camp argues that the first living organisms were likely genes, information-containing single molecules, *that could catalyze their own replication.* These simple, self-replicating molecules would then, under the influence of selective pressures, evolve into increasingly complex organisms and, on Earth at least, could have eventually evolved into organisms with modern Terrestrial biochemistry. The genes-first hypothesis has a significant advantage over the metabolism-first camp.

*The internet, with its complex network of servers, is similarly robust. In fact, the internet was originally built under funding from the U.S. Department of Defense as a communications network that could survive nuclear war. If any one node is knocked out, others are there to take its place.

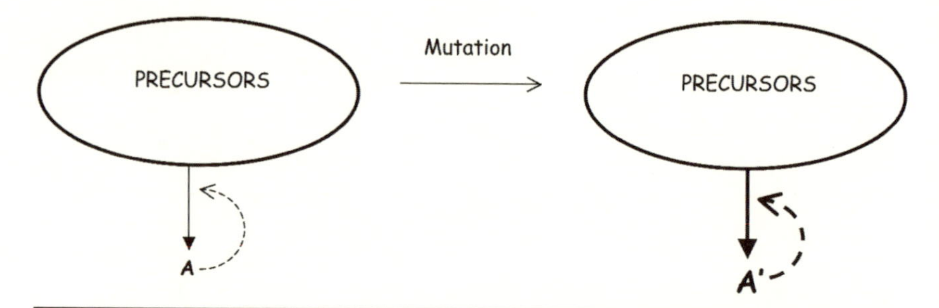

FIG. 5.4. The network problem is solved if there is a molecule, A, *that can catalyze the formation of itself.* In this case, a mutation to form a more fit molecule, A′—say, with better catalytic activity—would be an inheritable change. The RNA-world hypothesis theorizes that RNA is such a molecule.

Namely, while networks are fundamentally resistant to evolutionary change, a single molecule that catalyzes its own formation can evolve more readily *if modifications of the catalyst breed true* and can be passed down through the generations (fig. 5.4). The question, then, is: can we think up chemically plausible catalysts for which this holds? It seems we can.

Even the simplest self-replicating chemical systems, crystals, have been seen to exhibit the property of modifications breeding true. For example, irregularities in the surface of a crystal, such as the so-called screw dislocations that cause the surface of the growing crystal to spiral upward rather than to form as discrete layers, can continue to propagate as a crystal grows and, in some circumstances, can breed true if the crystal shatters and nucleates the formation of more crystals. Once again, though, we must remember that, even with mutations, replication alone is not sufficient to meet our definition of life; self-propagating modifications do not constitute evolution unless they provide a selective advantage for the "organism."

A. G. Cairns-Smith of the University of Glasgow has suggested, however, that some minerals have the right stuff to form "mineral genes" that can evolve and thus could have been the first living things. Clay is made up of sheets of charged silicates packed together in layers.* Cairns-Smith theorizes that the charges in one layer of clay could act as a template and catalyze the formation of a complementary new layer of clay—the "next generation" layer. Mistakes, which would be "mutations," occurring dur-

*The slickness of wet clay arises when water intercalated between the layers lubricates them and allows them to slide past one another.

ing copying would be inherited by all follow-on layers, and if these mistakes improved the efficiency of the replication process, they would provide a selective advantage, thus fulfilling our definition of life. These first *in*organic organisms, Cairns-Smith suggests, provided the scaffold on which life as we know it—built of sugar and spice and everything nice—later evolved. He notes (as we touched on in the previous chapter) that ions in clay can act as catalysts to speed up organic chemical reactions and, over the course of millions of years, could have been a significant source of RNA polymers and polypeptides in the primordial soup.

Sadly, though, Cairns-Smith's arguments suffer from some potentially serious weaknesses. Probably most critically, chemistry anything like that proposed by Cairns-Smith has never been demonstrated in the laboratory; it remains very much an open question whether defects and irregularities in clay mineral sheets can be made to "breed true." And with respect to the origins of life on our planet, it may be telling that no remnants of the clay-based metabolism exist in modern metabolism: there is nothing in our biochemistry that looks even remotely like silicate clay minerals. Cairns-Smith has argued that this simply means the original clay genes and catalysts have been completely and utterly replaced without leaving even the slightest vestigial traces. This is, of course, quite possible; clays are not the world's best catalysts and thus we might expect them to be completely abandoned after evolution invented better, organic catalysts. Still, were such vestiges present, they would be nice, tangible evidence supporting such origins. In their absence, we are left only with fascinating, but as yet experimentally unsupported, speculations.

What we really want is a relatively simple molecule that, unlike the hypothetical clay chemistry of Cairns-Smith, has been shown in the laboratory to be capable of, if not making more copies of itself, at least making molecules similar to itself. An example of such a molecule has been provided by Reza Ghadiri's group at the Scripps Research Institute in La Jolla, California. Ghadiri and his team designed a polypeptide (a very short protein) that autocatalytically directs the synthesis of copies of itself from two smaller, chemically activated polypeptide fragments, each consisting of half of the template sequence. Ghadiri himself, however, has argued that his polypeptides simply show that molecules that copy themselves from (slightly) simpler precursors are physically possible; they are poor candidates for the origins of life, however, because plausible prebiotic reactions are unlikely to generate appreciable amounts of the activated polypeptides that are the fodder of this reaction.

A similar problem has arisen in efforts to invoke DNA as the original self-replicating molecule. Starting in 1968, Leslie Orgel began to explore the idea that a single strand of DNA could serve as a template upon which activated nucleotides could spontaneously polymerize to form the complementary sequence. After decades of work, the conclusions are mixed. Polymerization can proceed reasonably efficiently starting from phosphorimidazolide monomers (an activated nucleotide we discussed in the previous chapter), but only on templates that are rich in the nucleobase cytosine. For example, the sequence GGCGG is obtained in 18% yield from the template CCGCC. But cytosine-poor polymers make very poor templates, and so the GGCGG product cannot be used to synthesize more of the CCGCC starting material. The reaction thus grinds to a halt after just one round and is not self-replicating.

Orgel's work with DNA-templated DNA polymerization and Ghadiri's work with self-replicating peptides raise another chicken-or-egg conundrum. Namely, in current Terrestrial biochemistry DNA encodes the information necessary to make proteins, and proteins are required in order to copy DNA. It is thus not at all clear whether either of these two species can replicate without the other, and thus it is not at all clear that either of these two could have been the first self-replicating molecule. A more promising candidate for the chemistry of the origins of life would be a molecule that has been shown to be able to, at least in a small way, copy itself from simpler, monomeric precursors. Just such a molecule may be at hand in the form of RNA.

The RNA World

As we hint above, by as early as the late 1960s, Francis Crick (co-discoverer of the structure of DNA), Leslie Orgel (we've met him before), and Carl Woese (whom we'll get to know better in chapter 7) had each independently developed the hypothesis that nucleic acids could have formed the basis of the first living things, a hypothesis that motivated much of Orgel's later work on DNA-templated DNA polymerization, described above. But underlying that work was the assumption that some inorganic catalyst would be responsible for the polymerization reaction. While the ability of RNA molecules to fold up into the sorts of intricate, three-dimensional structures associated with protein-based catalysts was well established by the early 1970s, the intellectual stance that proteins are *the* biological catalysts was so firmly entrenched—indeed, it was considered a major element of biology's *central dogma*—that the idea of nucleic acids as catalysts seemed far fetched; molecular biologists had relegated RNA to secondary, or more precisely, tertiary importance. In us, for example, RNA is neither the genetic material (a job carried out by DNA) nor the catalytic machinery by which metabolism is conducted (a job carried out by proteins). Instead, RNA was long thought to serve only lowly tasks such as messenger RNA's role in transporting genetic information from the DNA where it resides to the protein-synthesizing ribosome. That said, RNA was known to serve as the genetic material in a few, small viruses (the retroviruses, for example, such as the AIDS-causing HIV, encode their genetic information in RNA). If only RNA could catalyze reactions as well! Then we'd have what we were looking for: a molecule that could form the basis of both genetics and metabolism. And in 1981 just such catalytic activity was discovered, like so many things in science, essentially by accident.

The catalytic possibilities of RNA were discovered in the laboratory of Thomas Cech at the University of Colorado. Cech and his students were studying the mechanism by which messenger RNAs, the molecules that carry the instructions describing how to build a protein from the DNA to the ribosome, are "processed." Two years earlier it had been discovered that the information contained in messenger RNA is not continuous; the "coding" regions of the message are interrupted by apparently meaningless sequences called "introns" that must be spliced out of the immature RNA. Cech found that cellular extracts from his favorite study organism, the single-celled paramecium *Tetrahymena*, could carry out the splicing reaction in the test tube. When faced with such an observation, the natural thing for a biochemist to do is to try to "fractionate" the extract into its component parts in order to discover which protein in the system is responsible for the catalytic activity. The first fractionation experiment they attempted, using an extract from the nucleus of the cell, carried out the splicing reaction perfectly, suggesting that they were on the right track. But there was a fly in the ointment. The test tube "next door" to the nuclear extract was a "negative control" that lacked the extract, and yet splicing seemed to have occurred there as well.

At first they thought they had just mixed up the tubes, but repeating the experiment another *five* times produced the same puzzling result. Was the problem that the RNA they were using for the splicing reaction had been purified from *Tetrahymena* and thus was contaminated with a small amount of *Tetrahymena* proteins, including, perhaps, the splicing protein? Treatment of the RNA with proteases (enzymes that break down proteins) did not stop the reaction, suggesting that contaminants were not the source of the splicing reaction. Finally, in desperation, Cech's group synthesized the RNA in another organism, *Escherichia coli*, thus rendering the test RNA entirely free of *Tetrahymena* proteins. And still the RNA spliced. With such compelling evidence that the reaction could occur even in the absence of *Tetrahymena* proteins, Cech finally went public with a positively revolutionary idea: an RNA molecule could splice itself *without the help of proteins*. And while, in the strictest sense, the RNA splicing reaction is not catalysis—a true catalyst remains above the fray and is not modified in the chemical reaction it fosters—this result was the first hint that RNA could accelerate biochemical reactions in a manner previously thought solely the realm of proteins.

It immediately became apparent that Cech's autocatalytic RNA was not a one-off example. Sidney Altman, at Yale University, had simultaneously been studying another catalytic RNA-processing step: the maturation of transfer RNA (more about these small RNAs in the next chapter). One of the last steps in the process is cleavage of the RNA to remove several nucleotides from one end. The purified cellular component that carries out this process is a two-molecule complex consisting of a protein tightly bound to an RNA. Altman had been assuming that, as was dictated by the central dogma, the *protein* was the catalytic agent and the RNA played some structural or recognition role. But when Altman's students attempted to run the reaction with the pure RNA they found that, in the presence of a large quantity of magnesium ions (presumably to take the place of the protein in its role of neutralizing the negative charges on the RNA and allowing it to fold), the RNA alone was catalytic. For these discoveries, Cech and Altman shared the 1989 Nobel Prize in Chemistry.

In the twenty years since the first ribozymes,* short for "RNA enzymes," were discovered, the list of catalytic roles played by RNA has grown enormously (fig. 5.5). Naturally occurring ribozymes include such interesting surprises as the ribosome, the massive protein-and-RNA "factory" in which proteins are synthesized: proteins, it turns out, are synthesized by the catalytic activity of RNA, an important point to which we shall be returning in the next chapter. Meanwhile, the catalytic activity of RNA is far more diverse than indicated by these naturally occurring examples; many distinct catalytic functions have been "selected for" artificially in the laboratory. These additional catalytic activities include the cleavage of DNA-RNA hybrids, the cleavage of DNA, the ligation (sticking together) of two RNA molecules, the ligation of DNA, acyl transfer (the transfer of an organic group from a phosphate to another molecule), the cleavage of peptide bonds, the formation of peptide bonds (from activated, acylated compounds), and the insertion of a metal into a porphyrin to form heme, the organometallic compound at the heart of such diverse metabolic processes as photosynthesis (in the form of chlorophyll) and oxygen transport (in hemoglobin).

Do ribozymes provide a solution to the origins-of-life question? As

*Not to be confused with *ribosomes,* the protein synthesis machinery. We'll be discussing those in the next chapter.

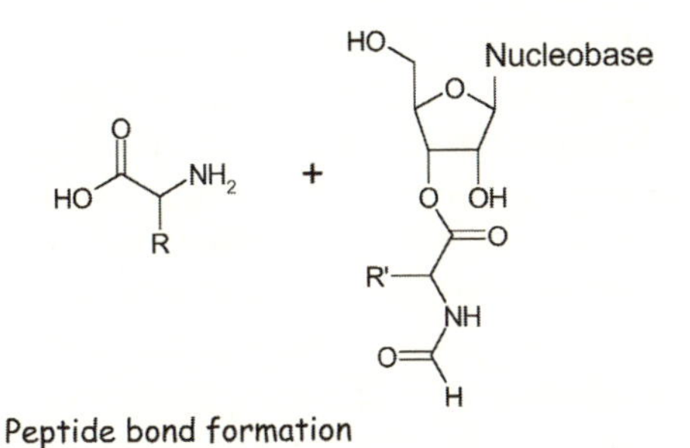

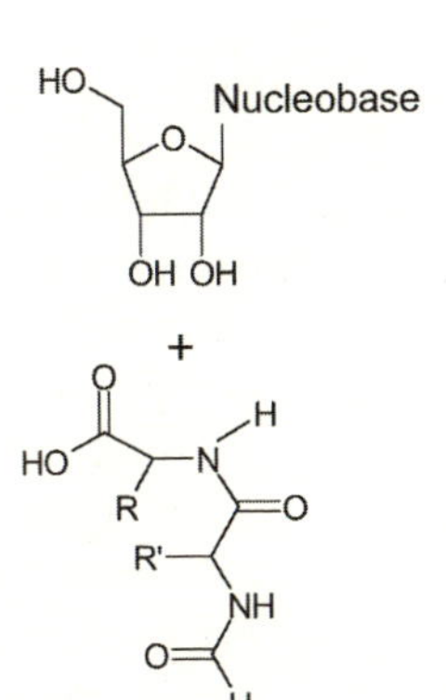

Peptide bond formation

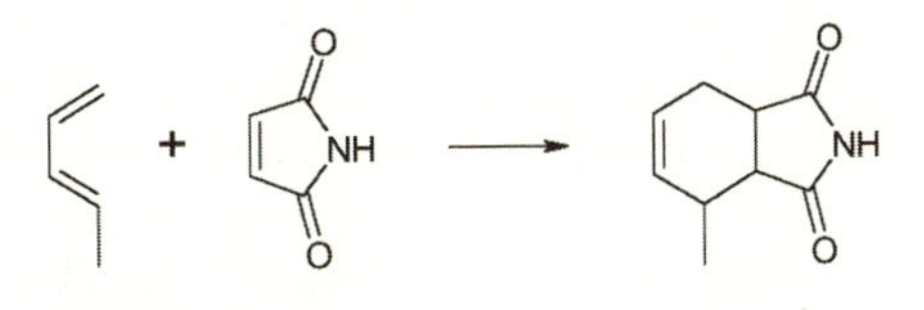

Carbon-carbon bond formation

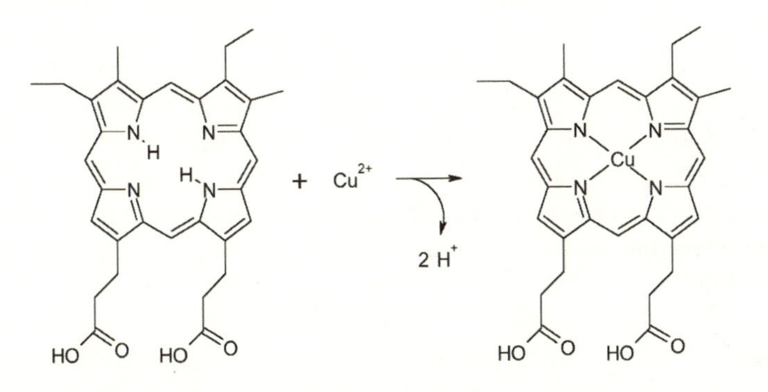

Porphyrin metallation

FIG. 5.5. While most ribozyme chemistry involves nucleic acid modifications (not a problem for prebiotic chemistry if the first organism required only that its ribozymes copied themselves), several artificial ribozymes have been reported that catalyze the modification of non-nucleotide substrates.

we have noted, it had long been known that RNA can serve as the repository of genetic information, and by the early 1980s it was established that RNA can serve as a catalyst as well. Could RNA, then, have been the original self-replicator? To serve this role requires a special kind of catalysis; a self-replicating molecule must be able to catalyze the formation of copies of itself. The formation of RNA (or other polymers) from simpler precursors is termed *polymerization.* The formation of a specific polymer sequence based on the sequence of another, templating polymer is called *template-directed polymerization.* Was the first living thing a ribozyme that carried out template-directed RNA polymerization? Is such a ribozyme even physically plausible?

To date, the best evidence affirming the latter question (which has, of course, implications for the former question) is work from David Bartel's laboratory at the Massachusetts Institute of Technology. Bartel's group employed a clever set of tricks to select ribozymes with polymerization activity from an enormous pool of partially randomized RNA sequences. To avoid searching through a huge morass of unfolded RNA polymers (no structure, therefore no function), Bartel's group started with a folded, catalytically active ribozyme sequence that performed ligation chemistry (the splicing together of two longer RNA polymers) and had exhibited some extremely limited polymerization activity. To this they added a short "priming sequence," which acted as the initiation site for polymerization, and a short RNA, which acted as an internal template that would guide the sequence-specific reaction. They then subjected this hybrid molecule to mutagenesis to produce a pool of 10^{15} different RNA sequences, some of which, they hoped, would exhibit the template-directed RNA polymerization activity they were looking for.

To fish out the presumably rare sequences with the desired activity, Bartel's group added activated nucleoside triphosphates to the pool of molecules, including the modified, sulfur-containing nucleotide 4-thioUTP. Any RNA molecule in the pool that could polymerize a short stretch of RNA on its template would incorporate this 4-thioUTP into its structure as 4-thioU. Bartel's students then separated out the 4-thioU-containing RNA molecules by forcing the RNA to migrate through a mercury-containing gel (mercury binds to the sulfur, impeding the motion of the 4-thioU-containing RNAs), converted the slowly migrating molecules into their complementary DNA sequences using a protein called reverse transcriptase, copied and amplified the pool of DNA sequences using the protein DNA polymerase, and converted the amplified pool of DNA sequences back into RNA using the protein RNA polymerase (phew!)—before starting the cycle over again. During the first few cycles they uncovered mostly mutant RNA that had "evolved" the property of covalently binding the modified 4-thioUTP without catalyzing template-directed polymerization. But with each additional round, the population of RNA molecules that exhibited polymerase activity (and thus incorporated many 4-thioU monomers) increased. Finally, after *eighteen* rounds of selection, copying, reengineering, and amplification, Bartel's students recovered several RNA molecules, 189 nucleotides in length, that could use a template to polymerize a specific, complementary sequence of RNA. They'd found molecules that, in theory at least, could catalyze the formation of copies of themselves.

But is Bartel's RNA polymerase ribozyme the self-replicating molecule we are looking for? It's close, but not quite there. The fidelity of the most efficient of the polymerizing sequences is only 97%; it makes three mistakes per hundred nucleotides polymerized. It is not clear whether this is high enough to ensure that a sufficient number of "offspring" are close enough in structure to the parent to maintain its polymerase activity. Worse still, the catalytic efficiency of even the best polymerase is depressingly poor. The best sequence reported to date extends its primer by only 14 nucleotides, which is obviously not nearly good enough for a 189-nucleotide molecule to copy itself (especially given that the catalytic ribozyme itself is relatively unstable and breaks down quickly compared with the rate at which it performs polymerization). That said, however, the ribozyme does have many of the characteristics we are looking for. It is quite *un*selective in its choice of templates and thus, in theory, it could copy a copy of itself (were it only about twenty times more active). And it does show the right type of activity—it synthesizes RNA polymers with linkages between the correct set of hydroxyl groups. Thus Bartel's work is a strong indication that sufficiently catalytic ribozyme RNA polymerase sequences can exist, and it provides a major boost for the idea at the core of the RNA-world hypothesis: that the first living thing on Earth was an RNA molecule that could (and did) copy itself.

While it's definitely premature to consider it established fact, the RNA-world hypothesis is probably the closest thing we have to a chemically plausible and experimentally supported theory of the origins of life. In RNA we have a molecule whose credentials both as a genetic material and as a catalyst are well established. And not just any type of catalysis, but RNA-templated RNA polymerase activity, the very type of activity that would be fundamental to the simplest living thing, were that living thing built of RNA. And while, admittedly, the best ribozyme RNA polymerase described to date can't copy itself, or even copy a copy of itself, this is presumably a quantitative issue and does not reflect a fundamental inability of RNA to perform such catalysis. Moreover, while it's not clear how RNA polymers might have been synthesized under prebi-

SIDEBAR 5.1

Weighing the Probabilities

Ribozyme sequences capable of catalyzing the template-dependent polymerization of new RNA seem to be rare. In order to create the first (and not a very efficient one at that), Bartel and coworkers had to start with an already folded, catalytic RNA (to cut down on search by minimizing the number of unfolded sequences investigated). And still they required eighteen rounds of selection and optimization—which couldn't happen on a prebiotic world (selection in the natural world requires an ability to *replicate* and these molecules aren't there yet)—to make a sequence that is somewhat active but *still can't copy itself;* the longest polymer synthesized by the most efficient of Bartel's ribozymes was just 14 nucleotides long, less than one-tenth the length of the ribozyme itself. And still Bartel's students had to test (in parallel, fortunately!) well in excess of 10^{15} sequences to find one that achieved even that level of activity. If more efficient polymerase sequences are out there they are presumably rarer still.

An illustration of the improbability of spontaneously generating a self-replicating RNA was provided to me (Plaxco) in the mid-1980s, when I was an undergraduate student touring potential grad schools. It was just a couple of years after the RNA world had first been postulated, and I was sitting in the office of Walter Gilbert, a Nobel Prize–winning molecular biologist from Harvard who was an early proponent of the theory. While we were chatting, a student interrupted; she popped her

otic conditions, at least we understand how most of the precursors of RNA were likely synthesized in the primordial soup. Lastly, as we explore in great depth in the next chapter, we can map out the route by which a simple, self-replicating RNA polymer could have evolved into us via a chemically plausible, stepwise evolution. The RNA world has a great deal going for it as an origins-of-life theory. But is the whole package now neatly wrapped up? In a word, no. (But what would be the fun of that, anyway?)

Several serious questions remain regarding the plausibility of the RNA world. One is the sheer improbability of spontaneously generating one of the rare sequences that could copy itself via the random, prebiotic polymerization of RNA monomers (see sidebar 5.1). This becomes even more of a hurdle when one realizes that the first living thing *had to consist of two such sequences together*! Why? Because it seems exceedingly unlikely that any molecule can *serve as its own template.* To serve as a template upon which a new RNA molecule can be synthesized, a molecule must be unfolded and exposed to the monomers that will polymerize on it. And unfolded molecules are not catalytic; catalysis is inti-

head in the door and simply said, "Nothing's come up yet." Gilbert looked at me thoughtfully, and said, "You look like you can keep a secret," and proceeded to describe the experiment. The student had made a test tube full of random RNA sequences, fed them with activated nucleotides, and, once a week, was checking to see whether any one length of sequence was replicating and thus increasing in number. Given that it's been almost twenty years since that chance encounter and Gilbert has never published the results of said experiment, I think it's a safe guess that none of the sequences ever started to dominate the mix (and it seems like the statute of limitations has run out on my promise not to tell this story). Fast forwarding fifteen years, Bartel's work (at neighboring MIT) suggested that any molecule in a test tube—or even a swimming pool—filled with fully random RNA sequences is extremely unlikely to exhibit polymerase activity. And, to make matters worse, both Gilbert's and Bartel's experiments started with chemically activated RNA precursors *of a single handedness.* How this homochirality is achieved and how activated nucleotides are synthesized prebiotically remain serious, unanswered questions, and thus the probability of randomly synthesizing a self-catalytic RNA-based RNA polymerase seems super-astronomically improbable.

Speculations such as these have led many to postulate that there was a "pre-RNA world." That is, that the first organisms were based on some other, simpler RNA-like polymer. Other than RNA, however, no other polymer has ever been demonstrated in the laboratory that exhibits template-directed polymerization, so these speculations remain just that, speculations.

mately related to the precise, three-dimensional placement of atoms. Thus, the first living chemical system did not arise until a ribozyme template-dependent RNA polymerase sequence was spontaneously generated in the presence of a second sequence that encoded the same catalytic function (such as a copy of itself, although any sequence that worked as a polymerase would be suitable) and could serve as a template. This need for two sequences *squares* the already significant *improbability* of the random synthesis event.

And as if this were not enough, there is yet another chemical effect that reduces the probability of forming a properly catalytic RNA. Sugars (such as the ribose in RNA), amino acids, and many other organic compounds are *chiral.* That is, just as your left hand cannot be superimposed on your right hand no matter how much you turn and twist it, these molecules are not superimposable on their mirror image. Terrestrially, we find that proteins consist entirely of "left-handed" amino acids. Such *homochirality* is critical; polymers made up of mixed-handed amino acids are very difficult to fold into unique, functional, three-dimensional structures. Ribose is similarly chiral, and it is thought to be

SIDEBAR 5.2

The Origins of Homochirality

Louis Pasteur's career was marked by far more than his debunking of spontaneous generation. He invented "pasteurization," for example. He was also the first person to note, some fifteen years before his work on "corpuscles that exist in the atmosphere," that the component chemicals of life are *chiral,* that is, they are not superimposable on their mirror image. Your hands are a convenient example of chirality; they are mirror images of one another that are not superimposable: no matter how many ways you twist and turn it, your left glove will not fit on your right hand. Amino acids and sugars are similarly chiral, coming in left- and right-handed versions that, while similar in many aspects, are not interchangeable.

On Earth, proteins are composed entirely of "left-handed" amino acids, termed L-amino acids. And nucleic acids—DNA and RNA—contain only "right-handed" sugars termed D-(deoxy)ribose. (We note, however, that as used by chemists, the terms *right* and *left* are, in effect, arbitrary, historical designations and do not imply that the handedness of amino acids and ribose are opposite one another.) We understand why this *homochirality* must occur; as mentioned in the text, polymers of mixed-chirality monomers tend not to fold, and thus homochirality provides a selective advantage for organisms that need to fold their proteins and RNA into functional, active forms. But is there a selective advantage associated with L-amino acids and D-ribose versus D-amino acids or L-ribose?

Historically, the answer to this question was "probably not." The reason is that the chemistries of mirror-image molecular pairs (termed *enantiomers*) are indistinguishable. This speculation, though, was put onto a firmer experimental footing in the mid-1980s by Stephen White, then at Caltech. One of the world's experts in the chemical synthesis of proteins, White undertook the task of synthesizing a protease (a protein that catalytically cleaves other proteins) made up entirely of D-amino acids. When synthesized, the protease folded just fine, but as expected it *folded into the mirror image of the naturally occurring protein.* This mirror-image protein was just as catalytically active as the naturally occurring protein, but, consistent with its mirror symmetry, only in cleaving protein chains consisting of D-amino acids. Just as your right-hand glove fits only your right hand, this right-handed protein binds to and cleaves only right-handed substrates. Polymers of *mixed* handedness do not work well. Polymers in which the chirality of the monomers changes randomly from left to right tend not to fold (at least not into nice, regular structures), and without folding, there is little in the way of catalysis. Thus homochirality (all monomers of one handedness) is probably a universal property of living things.

While there is a strong selective advantage associated with homochirality, the fact that a protein consisting of D-amino acids is just as catalytically active as its mirror-image partner strongly suggests that there is no selective advantage associated with the occurrence of

L-amino rather than D-amino acids in Terrestrial biochemistry. Similar arguments can be made for the use of D-ribose in RNA and D-deoxyribose in DNA. So why, then, were these two "hands" selected for on Earth, and not their mirror images? Two broad classes of theories have been suggested in an attempt to address this question. The first is that the selection of L-amino acids and D-ribose was biological: it may simply be an accident of our evolutionary history. That is, while there is strong selective pressure to adopt one handedness, which of the two hands selected was a simple, random choice that became frozen in place by selective pressures. Alternatively, the current handedness of life on Earth may be due to abiological processes that preferentially degraded one handedness, leaving organisms that utilized the other handedness at a selective advantage.

The abiological processes postulated to produce such excesses include the action of circularly polarized light on the presolar nebula, asymmetric effects in the weak nuclear force, and, perhaps most implausibly, coupling between the Earth's orbit, its spin, and its prebiotic chemistry. The latter theory, proposed by the Chinese scientists Y. I. He, F. Qui, and S. D. Qi, is probably easiest to dismiss; there is no known mechanism by which the direction of the Earth's orbit and spin (which, when coupled, are not superimposable on their mirror image and thus are, admittedly, chiral) could couple in such a way as to affect chemistry. With no chemically plausible explanation for how this might happen, and no—even unexplained—laboratory demonstration of such an effect, we are probably safe in ignoring this theory.

Weak nuclear force effects are probably likewise not a very plausible theory. In the 1950s it was discovered that β-particles, energetic electrons emitted from atomic nuclei during some forms of radioactive decay, are preferentially emitted with a specific handedness (in terms of their direction of travel and direction of spin). This implies that the weak nuclear force, the force that holds nuclei together, is itself asymmetric, an effect that could, potentially, have implications in the origins of homochirality. Despite intensive investigations, however, no one has demonstrated the selective degradation of one handedness over another through the effects of β-radiation, thus its role in prebiotic chemistry must be seriously questioned.

The theory suggesting that circularly polarized light produced chiral excesses on the early Earth is on perhaps the best footing, but even this one is only very weakly supported by laboratory experiments. Like any charged object moving through a magnetic field, an electron traveling through the intense magnetic fields of a neutron star will spiral. In tracing out a spiral, the electron is accelerating, and Maxwell's equations (a set of equations describing the properties of electromagnetic fields) tell us that an accelerating charge will emit light. Given the typical energies of electrons under these circumstances, much of the light will be emitted as UV and, because the electrons are spiraling, this light will be circularly polarized (i.e., the electric field vector that describes the light will oscillate in a circular fashion, as opposed to the more well-known polarization caused by your

Sidebar 5.2 continued

sunglasses, in which the electric field vector of the polarized light is constrained within a plane). The two enantiomers of a chiral molecule absorb circularly polarized light in different amounts.

Based on these observations, it has been postulated that if (1) there was a nearby neutron star producing copious amounts of circularly polarized UV light and (2) the prebiotic precursors to life arose in the presolar nebula (where they would receive the full brunt of any UV light), then this might account for the selection of L-amino acids and D-ribose on Earth. Laboratory studies of this mechanism, however, indicate that it is a very weak effect. When amino acids were blasted with so much UV that 99% of all molecules (of both handedness) were destroyed, the remaining dregs of amino acids showed enrichment of at most only a few percent. If such an effect is a mandatory step in the formation of life, then the origins of life were a fortuitous event indeed.

equally critical that, in order to fold, a functional ribozyme must similarly be homochiral. The requirement for homochirality represents a potentially serious problem for the RNA world. The problem is that biological processes, such as the prebiotic chemistry of the early Earth, produce equal amounts of both "left-" and "right-handed" molecules (see sidebar 5.2). Thus it is highly improbable that random chemistry would produce a polymer molecule that contained monomers of only one handedness. To be precise, the probability of achieving homochirality in a 189-unit polymer polymerized from an equal-molar mixture of left- and right-handed monomers is 1 in 2^{189} (1 in 8×10^{56})! This number of 189-nucleotide RNA sequences would, taken together, be as massive as forty thousand Suns, which nicely illuminates the extent to which the requirement for homochirality increases the improbability of a self-replicating RNA sequence arising spontaneously from an activated mixture of ribonucleotides.

The requirement for homochirality—and the poor efficiency with which RNA is synthesized under plausible prebiotic conditions—reduces the probability that RNA-world-type chemistry is at the root of the origins of life. But does this render the RNA-world hypothesis invalid? It does not. For one, the need for homochirality does not push the probability of spontaneously generating a self-copying ribozyme to zero, and the anthropic principle says that even super-astronomically improbable mechanisms may lie at the heart of the origins of life—our existence says only that the probability of life arising in our Universe is not zero, but it could be infinitesimally close to zero. Furthermore, some of the inorganic RNA-forming catalysts that we described in the previous chapter could, conceivably, be *stereoselective.* That is, they could produce polymers of either pure left- or pure right-handed ribonucleotides (they would presumably produce equal amounts of left- and right-handed polymers, but we don't care about this; we care only that each individual polymer molecule is homochiral). To the best of our knowledge, though, this issue has not been investigated. (If any of you gentle readers are looking for a good Ph.D. thesis project . . .) Alternatively, there may have been a "pre-RNA world" in which a nonchiral, RNA-like polymer was the basis of the first life, and chiral RNA came into the picture only later, under the influence of selective pressures.

Several such potential pre-RNA polymers have been implicated to date. For example, Albert Eschenmoser, now at the Scripps Research Institute, has demonstrated that nucleotides containing the sugar threose can be strung together into RNA-like polymers that form complementary duplexes, and has argued that threose, being a four-carbon sugar, is likely to be synthesized in greater yield than the five-carbon ribose under prebiotic conditions. Threose, however, is also chiral, and thus Eschenmoser's TNAs (threose nucleic acids) do not solve the homochirality problem. In contrast, Pernilla Wittung, then working in Peter Nielson's laboratory in Copenhagen, demonstrated in the early 1990s that a nucleic acid composed of a nonchiral polypeptide-like backbone (peptide nucleic acids, or PNAs, consisting of nucleobases linked together via the molecule *N*-(2-aminoethyl)glycine, which is produced in the Miller-Urey experiment) can couple with itself to form stable double helices. Unfortunately, however, neither TNA, nor PNA, nor any other RNA-like polymer has yet been shown to possess catalytic activity.

Conclusions

A hundred and fifty years ago, Pasteur won the French Academy's prize for showing us how life *does not start.* And now, almost a century and a half on, we still do not know how it does. Self-replicating metabolic pathways? Self-replicating clays? An RNA molecule (or an RNA-like molecule) that can copy a copy of itself? We simply do not know, because every one of these theories faces major, unsolved hurdles. But at least in the RNA-world hypothesis we have a chemically and biologically plausible—if perhaps improbable—theory as to our origins.

Further Reading

General reading on theories of the origins of life. Brack, Andre (ed.). *The Molecular Origins of Life.* Cambridge: Cambridge University Press, 1998; Davies, Paul. *The Fifth Miracle.* New York: Touchstone, 1999.

Metabolism-first hypothesis. Orgel, Leslie E. "Self-organizing biochemical cycles." *Proceedings of the National Academy of Sciences USA* 97 (2000): 12503–7.

The Fe-S world. Cody, G. D. "Transition metal sulfides and the origins of metabolism." *Annual Review of Earth and Planetary Sciences* 32 (2004): 569–99.

The lipid world. Segre, D., Ben-Eli, D., Deamer, D. W., and Lancet, D. "The lipid world." *Origins of Life and Evolution of the Biosphere* 31 (2001): 119–45.

The RNA world. Gesteland, R. F., Cech, T. R., and Atkins, J. F. (eds.). *The RNA World.* Woodbury, NY: Cold Spring Harbor Laboratory Press, 2005.

RNA-polymerase ribozyme. Johnston, W. K, Unrau, P. J., Lawrence, M. S., Glasner, M. E., and Bartel, D. P. "RNA-catalyzed RNA polymerization: accurate and general RNA-templated primer extension." *Science* 292 (2001): 1319–25.

CHAPTER 6

From Molecules to Cells

After taking part in the unraveling of the DNA double helix, Francis Crick moved to the logical next step of investigating the expression of genetic information: how the sequence of nucleotides so neatly lined up in the double helix structure is eventually translated into amino acid sequences. With his wildly creative thinking he contributed some insights that turned out to be true—such as the adapter hypothesis, which essentially predicted the role of transfer RNAs—and others that were perhaps just that little bit too imaginative, like the "comma-less genetic code," which restricted the use of three-letter codons to those that would be safe against one-letter shifts of the reading frame (we describe the genetic code below). He was much intrigued both by the complexity of the protein biosynthesis machine and by the universality of its language.

In 1972, Crick resorted to yet another wild idea to explain the puzzling phenomenon that all the millions of species on our planet use essentially the same incredibly complex protein synthesis machinery, with the same genetic code. In a paper he coauthored with Leslie Orgel, which appeared in the planetary science journal *Icarus* the following year, the two argued that, in particular, the protein synthesis machinery is so complex that it was hard to understand how it could have been created by the slow, stepwise progression of evolution. This quandary is so serious, they proposed, that we must consider even the wildest alternative hypotheses. For example, they argued, a possible alternative solution would be that life on Earth, with its seemingly impossibly complex biochemistry, may have been artificially designed by intelligent aliens from some other planetary system in our galaxy, who then purposefully sent it here by space mail. How did the aliens in question solve the problem of the difficulties associated with the origins of our complex biochemistry? They, went the theory, had a much simpler biochemistry. Simple enough, Crick and Orgel argued, that it, unlike ours, could have arisen by chance. The two further decorated their proposal—which builds on the earlier panspermia ideas, discussed in chapter 5—with plenty of detail, including the design of the spaceship that might have delivered the first spores to Earth. It is a remarkable tribute to the complexity of contemporary Terrestrial biochemistry that, after decades spent deciphering some of its greatest mysteries, these two highly distinguished researchers still found many aspects of biochemistry so puzzling that they considered such a radical alternative seriously (or at least semi-seriously: Orgel admits that the paper was for him rather tongue-in-cheek, but he suspects "Francis took it somewhat more seriously").

Today, more than three decades onward, few researchers would resort to such speculative explanations, but the evolution of protein biosynthesis still poses a serious conundrum: protein synthesis as it occurs today in all cellular organisms (and even, separately, in certain subcellular compartments, the mitochondria) requires an incredibly large toolkit, including ribosomes consisting of more than fifty proteins and at least three large strands of RNA, tRNAs (the *t* stands for "transfer"; this RNA's job is to transfer an amino acid to the growing polypeptide chain), and dozens of different nonribosomal proteins, including the tRNA synthetases (which covalently attach the appropriate amino acid to each tRNA) as well as initiation, elongation, and release factors. If any one piece is removed, the whole process might fail. How could this have arisen by the stepwise, random workings of evolution? How could the RNA world transform itself into a DNA-RNA-protein world? (And if we rewound and ran the film again, would the story unfold the same way? See sidebar 6.1.) Luckily, clues to this mystery are contained within the very complexity of the protein synthesis machinery itself.

My Name Is LUCA

From the extraordinary degree to which the biochemistry of all Terrestrial life is similar, it seems that everything living on Earth today is related through some long-lost great-to-the-*n*th-power grandmother, the Last Universal Common Ancestor (LUCA) of all living things on Earth. Arguing in reverse, we can look across all life on Earth to identify those things we share in common, under the economical assumption that any traits that are common across all life are shared because we all inherited them from LUCA, and thus she shared them too. Using this comparative approach, biochemists have been able to infer quite a bit about LUCA and her biochemistry. For example, we know that LUCA stored her genetic information in DNA, that she possessed a couple of hundred proteins working as enzymes, receptors, or transporters, and that she made use of the twenty amino acids that are now standard in proteins. Similarly, LUCA translated the information in her genes to make functional proteins using the same complex, RNA-based machinery that is in use today, and the "genetic code" that LUCA used to translate DNA

SIDEBAR 6.1

Was There Another Way?

All life on Earth today employs proteins for the vast majority of its catalytic functions. Similarly, save for a few small, simple viruses, all Terrestrial life employs DNA as its genetic material. But did this have to be? That is, is there some physical or chemical imperative that forced evolution's hand to make these "choices"? Or was the choice to adopt protein catalysts and DNA information storage simply a historical accident? Can we even meaningfully speculate on these issues?

As we describe later in this chapter, the "invention" of DNA was encouraged by the selective advantage that DNA provides as an information storage agent in terms of its chemical stability. But stability is not the sole requirement of a genetic material. It must also be able to mutate (sometimes, but not too often!) to create new information (i.e., to allow evolution to proceed), without changing its physical properties so much that the replication and translation machineries no longer recognize it. This principle, termed COSMIC-LOPER by Steve Benner of the University of Florida (*C*apable *o*f *S*earching *M*utation-Space *I*ndependent of *C*oncern over *L*oss *o*f *P*roperties *E*ssential for *R*eplication), states that an ideal genetic material will be one whose physical properties remain largely unchanged by changes in sequence. And in addition to being chemically stable, DNA is COSMIC-LOPER; the shape and chemistry of the double helix is, to a first approximation, independent of its nucleotide sequence.

But was DNA the only way for evolution to meet the demanding, simultaneous requirements of chemical stability and COSMIC-LOPER? There are other nucleobase-containing polymers that form complementary helices (such as the PNAs and TNAs we discussed in chapter 5), which suggests there are other ways of storing genetic information in molecules. But do these other polymers have the right set of properties? Quite simply, we do not know.

And what about proteins? As we have also described in the text (chapter 5), the improved catalytic ability of protein-based enzymes relative to RNA-based ribozymes provides a strong selective advantage to any organism that evolves the ability to make proteins. But did these new-found catalysts have to be polymers of α-amino acids? The answer is a qualified "perhaps so." The arguments are severalfold. First, α-amino acids are formed in relatively high yield in the Miller-Urey experiment and are a relatively important component of the organic material in carbonaceous chondrites. It seems reasonable to assume that the ready availability of α-amino acid precursors would provide a significant boost for any organism using proteins, as opposed to some other polymer, to accelerate its metabolism.

A second issue is the physical chemistry of polypeptides, which seems to ideally suit them for folding into well-defined—and thus functional—structures. The "peptide linkage" between amino acids in a protein contains a hydrogen atom attached to a nitrogen atom (see sidebar 4.1). Because the nitrogen is strongly electronegative (has a high affinity for electrons), the hydrogen atom takes on a small (partial) positive charge. The peptide unit also contains a strongly electronegative oxygen, which takes on a partial negative charge. These two atoms can participate in a "hydrogen bond," in which a positively charged hydrogen effectively sits on top of a negatively charged oxygen. When a protein is unfolded in water, the hydrogen can hydrogen-bond to oxygen atoms in the surrounding solvent molecules, and the oxygen can hydrogen-bond to hydrogens in these same solvent molecules. When a protein folds, however, the solvent is excluded and thus all of the hydrogen bonds need to be formed internally.

As Linus Pauling (1901–94) realized back in 1948 (while sick in bed and doodling on some paper), polypeptides can satisfy all of their hydrogen-bonding needs internally. For example, they can fold into what he named an α helix, in which the hydrogen of one peptide linkage is hydrogen-bonded to the oxygen of a peptide linkage some three amino acids farther along the chain. Moreover, all of the bond lengths and bond angles in the amino acids are extremely stable in this configuration; there are no nasty clashes of one atom against another, and the bonds are all satisfied with the angles they have to adopt—thus researchers were not surprised to find that α helices are indeed a common feature of folded proteins when the first such structures became available some fifteen years later. More recently, Jayanth Banavar of Penn State University and Amos Maritan of Padova University in Italy have pointed out that the packing density of this helix is effectively perfect. If you want to twist a linear tube into a helix shape leaving as little empty space as possible, the radius of curvature of the helix has to equal the tube radius, and the pitch of the helix has to be precisely 2.512 times its radius. And for the α helices found in proteins, the ratio of the pitch to the radius is within a few percent of this value. This near-perfect packing also encourages the formation of a stable, folded state—because nature abhors voids.

While the existence of ribozymes proves that proteins are not the only polymeric catalysts, the one-two-three punch of proteins' greatly improved catalytic properties, their formation from readily available precursors, and their folding-friendly physical chemistry renders them extremely well suited for the role of metabolic catalysts. So we really must wonder, when we finally travel to Alpha Centauri and meet up with the Centaurians, will they, too, use polymers of α-amino acids as their dominant catalysts? To quote Banavar, "Let us endeavor to do so and find out."

Benner, S. A. "Chance and necessity in biomolecular chemistry: is life as we know it universal?" In *Signs of Life: A Report Based on the April 2000 Workshop on Life Detection Techniques,* ed. Committee on the Origins and Evolution of Life, National Research Council. Washington, DC: National Academies Press, 2002.

Maritan, A., Micheletti, C., Trovato, A., and Banavar, J. R. "Optimal shapes of compact strings." *Nature* 406 (2000): 287–90.

sequences into the sequences of amino acids in proteins was identical to the code used in our cells.

LUCA's biochemistry was thus already fairly complex; by the time LUCA was around, life had come a long way since the birth of the first simple, self-replicating molecules. But because so much was lost in going through the bottleneck that was LUCA and, presumably, many earlier bottlenecks (we have to remember that there may have been thousands of other species, contemporary to LUCA, whose descendants just weren't lucky enough or fit enough to survive in competition with her and her offspring), we can only speculate about how the first cells came into being, and which of the many compounds and functions that we now consider essential may have been absent in LUCA's progenitors. But we do have constraints that can guide our speculations: we assume that the primordial soup was the starting point and LUCA was the final product.

From RNA to LUCA

One of the most widely accepted theories concerning the phase between primordial soup and ancestral cells is the RNA-world hypothesis, which we introduced in the previous chapter. The discovery of ribozymes as potential relics of the RNA world was a major inspiration for RNA biochemists, as it suggested that RNA could carry out the various functions considered essential for life. But the first ribozymes to be discovered were limited to the processing of other RNAs or, perhaps worse, to the processing of themselves (i.e., as we mentioned in chapter 5, they aren't really catalysts in the true sense of the word). This is hardly the sort of diverse chemistry around which complex life can be built; for that we'd expect to need, at the very least, reactions that convert small molecules that look nothing like RNA into the true precursors of RNA.

The first ribozyme-catalyzed reaction that does *not* involve the processing of RNA was observed by Thomas Cech's group in 1992 with a genetically modified *Tetrahymena* ribozyme. The altered ribozyme could cleave the ester bond between the amino acid methionine and its matching tRNA, so it catalyzes the reverse of the reaction catalyzed by tRNA synthetases, the protein-based catalysts that add amino acids to tRNAs as the first step in the synthesis of proteins. Three years later, Michael Yarus, a colleague of Cech's at the University of Colorado, and his team screened a random mixture of 10^{14} different RNA sequences and managed to select a sequence that catalyzes the attachment of an activated amino acid to itself to form an "aminoacyl" group (see fig. 5.5). More recently, Peter Lohse and Jack Szostak at Harvard Medical School have demonstrated an RNA sequence that is able to transfer aminoacyl groups to a free amino acid to form a dipeptide, just as the peptidyltransferase center of the ribosome does when it elongates the polypeptide chain by one unit.

It thus seems that RNA can bind substrates and catalyze a range of reactions above and beyond the postulated ability of some RNA molecules to copy or edit themselves (as they might have done if the RNA world was not just an intermediate but the beginning of life). This chemical promiscuity might have made all the difference to the earliest organisms. The first self-replicating, evolving, hence living, thing would have been the only organism in the history of the planet that was not in competition with other organisms. That situation would have changed just a soon as it produced daughters. Immediately, competition is set up for the precursors these organisms need to reproduce, and all too soon some reagent or another would have become hard to obtain. Then what? Then death. Unless, of course, one of these replicating RNA strands was error prone (or, put more positively, could evolve) and started to copy some different RNA sequence that catalyzed some reaction not directly tied up in replication but that instead converted some not-quite-right precursor into the correct, but now rare, precursor. This would provide a *tremendous* selective advantage for this organism. It would also mark the invention of *metabolism.*

It is possible that complex metabolisms arose during the era of RNA-based organisms. The present-day widespread use of ribonucleotides as cofactors in key metabolic pathways suggests that these pathways may have originated during the RNA era. Examples include the use of the ribonucleotide ATP as the main energy "currency" of the cell, and the use of the ribonucleotides $FADH_2$ and NADH as the molecules that accept and donate reducing potential (remember: reduction potential is the ability to give up electrons in a chemical reaction). These putative remnants of the RNA world suggest that RNA-based organisms might have had a rich metabolism that included complex oxidation and reduction chemistries. And since the organism was enclosed by something (possibly even a cell membrane), all this metabolic machinery, like the replicative machinery that preceded it, was physically associated with the genes that encoded it, providing them and them alone with the all important selective advantage.

But if our ancient ancestors had such rich RNA-based metabolisms, why don't we see RNA-based metabolisms today? The presumed answer is that any RNA-based organism would have been driven out of the market, remorselessly outcompeted by its more fit cousins who had had the good fortune to invent or inherit protein-based catalysts. This is not surprising, as the catalytic efficiency of protein enzymes is typically orders of magnitude higher than that of comparable ribozymes. For example, while the so-called hammer-head ribozyme, which catalyzes the degradation of RNA, accelerates the degradation reaction a millionfold over the rate of the uncatalyzed reaction, a similar-sized, protein-based enzyme called RNase, which catalyzes the same reaction, is 100,000 times more active still. Proteins, consisting of twenty different types of amino acids (as compared with RNA's four nucleobases), are much more complex and thus, not surprisingly, generally much more catalytically flexible and efficient than ribozymes. Thus any RNA-based organism that learned how to synthesize proteins was at a tremendous advantage relative to its cohorts.

Of course, seeing a comparison between the exceptional catalytic ability of proteins and the rather paltry skills of RNA-based catalysts, skeptics might even question whether the RNA world ever existed. There was always the possibility of an alternative explanation: the relatively rare ribozymes might be fairly recent, and not very successful, experiments of evolution, rather than survivors of a principle that had once been more generally applicable. This ribozymes-first versus ribozymes-last argument raged through much of the 1990s (among those of us who cared), until it was finally resolved in favor of the primordial role of ribozymes. The resolution came with the unraveling of the mysteries of the ribosome, the subcellular factory that churns out the cell's proteins.

Even the prokaryotic (bacterial) ribosome (which is somewhat simpler than ours) consists of up to fifty-seven proteins and three RNA molecules folded together into an enormous macromolecular machine of several hundred thousand atoms in total (fig. 6.1). But which of these many components is responsible for the fundamental chemical reaction that the ribosome catalyzes, the synthesis of polypeptides? Long after the initial characterization of the ribosome it was assumed that this catalyst must be one of the proteins (remember: until 1982 it was believed that all biocatalysis was carried out by proteins). Research throughout the 1980s and 1990s, though, had shown that each individual ribosomal protein was expendable—that is, after removal of each of the many ribosomal proteins in turn, the ribosome continued to function. But despite several well-publicized—and quite possibly successful—attempts, nobody had unequivocally succeeded in constructing a completely protein-free version of the ribosome that remained active. Thus, the RNA-world community had to wait for an atomic-resolution structure of the ribosome to find out whether the ribosome is, at heart, a protein-based enzyme or an RNA-based ribozyme.

When a string of high-resolution structures of ribosomal subunits and complete ribosomes started to appear in 2000, the answer was per-

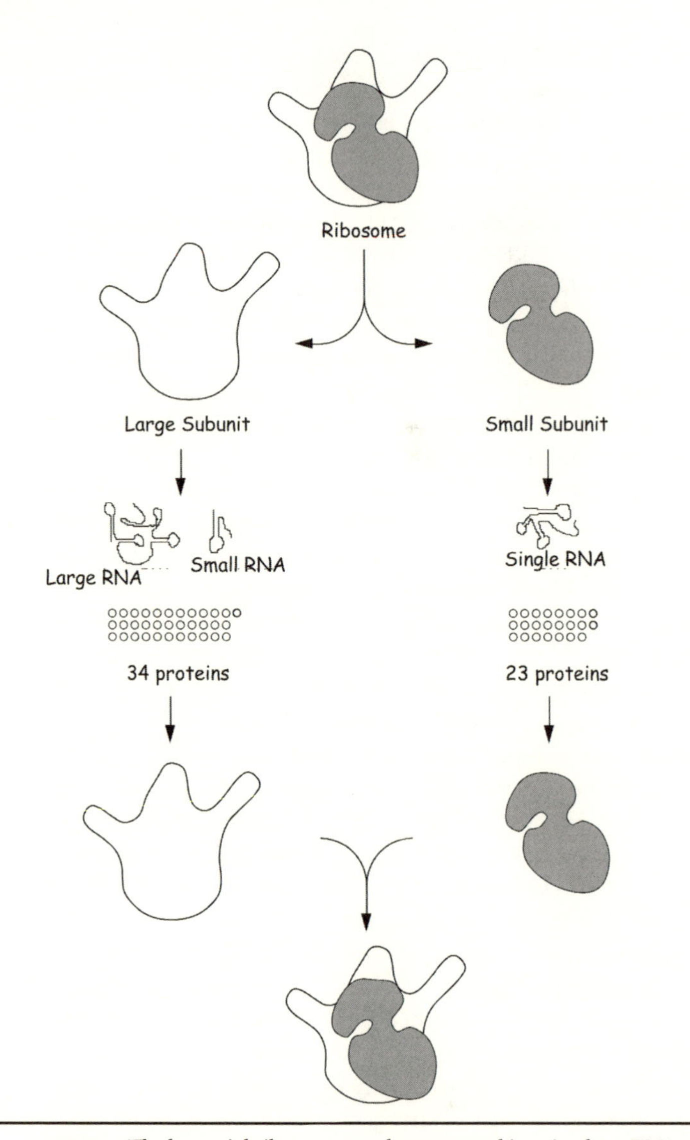

FIG. 6.1. The bacterial ribosome can be separated into its three RNA strands and up to fifty-seven protein molecules and then reconstituted to biological activity from these parts. Using this approach, researchers have found that the resulting ribosomes are still active if one of the proteins—no matter which one—is left out during reconstitution, showing that none of the proteins is crucial for the ribosome's catalytic function: the synthesis of polypeptides.

haps even clearer than expected. Reported by the groups of crystallographer Tom Steitz and ribosomologist Peter Moore, the structure of the larger of the two subunits of the ribosome from *Haloarcula marismortui* made the cover of *Science* magazine in August 2000. It reveals beautifully the peptidyltransferase site of the ribosome, the site at which each new amino acid is linked to the growing polypeptide chain in a mechanism that essentially reverses the polypeptide-cleaving function of a very well-characterized group of enzymes, the serine proteases. What must have delighted the RNA-world community more than any amount of mechanistic detail, however, was the observation that there is no trace of protein anywhere near the active site. The nearest protein is 1.8 nanometers (i.e., around 15 bond lengths) away and thus clearly excluded from any participation in the catalytic activity. As RNA champion Tom Cech concluded in a commentary accompanying the structure paper: "The ribosome is a ribozyme." It is now abundantly clear that the original catalyst for the synthesis of proteins was an RNA molecule, which was later surrounded by proteins that serve various supporting roles.

Polypeptides Join the Fold

Proteins are long polymers of specific sequence. The question thus naturally arises as to how such complex entities could arise in one fell swoop. In short, they probably could not. Instead, it is thought that the first polypeptides were short, random-sequence polymers of amino acids. But random polypeptides generally do not fold and are not catalytically active, and thus they lack the selective advantages we generally associate with true proteins. Given this, why would some previously peptide-less life form evolve the ability to polymerize amino acids? The chemical strength of the peptide bond implies that even simple peptides could form stable structural elements for an early cell. Similarly, homogeneous polymers of positively charged amino acids could help counteract the negatively charged phosphates in RNA and improve its ability to fold into functional structures (remember the more than fifty proteins in the ribosome, all of which are there to help the catalytic RNA fold). Thus, it seems, even random polypeptide sequences might have provided a selective advantage for the first organism to break out of the RNA-only mold. Moreover, random polypeptides could also have given evolution a boost simply by providing a new set of materials for it to work with.

The introduction of polypeptides into biology presumably evolved via a very close association of amino acids with RNA, which over hundreds of millions of years culminated in the complex and well-conserved mechanisms we still observe in today's protein biosynthesis. The first step of this evolutionary process might have been the use of amino acids or short peptides to add functionality to RNA. Evolution could easily have invented an RNA molecule that "charged itself"—that is, that autocatalytically added an amino acid to itself to form a covalent complex. As we described above, such molecules have been created in the laboratory (see fig. 5.5), so we know that RNA is capable of performing such chemistry. These mixed molecules could have provided a selective advantage for an originally RNA-only organism, say, by being more strongly catalytic due to the addition of a new functional group. Similarly, even in contemporary Terrestrial biochemistry we see examples of "aminoacylated" RNAs that are used to "donate" amino acids in metabolic processes.* Thus the first self-charging RNA could have resulted from a need to introduce amino acids (from Miller-Urey chemistry) into metabolic pathways aimed at making more RNA. Either way, it is easy to understand how the formation of the first RNA–amino acid complexes could have provided a selective advantage for RNA-based organisms. And if it existed, it seems likely that this amino-acid-charged RNA might have been the progenitor of today's tRNA (but without the need for today's tRNA-charging enzymes).

The second step in the evolution of protein synthesis was probably the formation of a ribozyme that could use the RNA–amino acid complexes to synthesize homogeneous or random polymers of amino acids. Once again, ribozymes capable of synthesizing peptide bonds starting from "activated" amino acids (amino acids participating in high-energy bonds, such as those in aminoacylated RNAs) have been demonstrated in the laboratory, so we know that RNA is capable of performing this sort of chemistry. But is such a step a plausible intermediate in the origins and evolution of life? The product of this reaction, short random or homogeneous polymers of amino acids, may have provided selective advantage when employed for structural roles in the cell because of their suitability. Under the influence of this selective pressure, evolution could have quite readily produced a ribozyme capable of synthesizing polypeptides from charged tRNA precursors. This ribozyme would have been the ancestor of the modern ribosome.

Random-sequence polypeptides, though, are not proteins. How did

*For example, a charged glutamyl-tRNA is used in the synthesis of porphyrins, chemistry that dates back to LUCA. And even the amino acid synthetases, the enzymes that now catalyze the acylation of tRNAs, use energy in the ribonucleotide ATP (adenosine triphosphate) to create the high-energy bond between the tRNA and the appropriate amino acid. Moreover, it is the ATP that carries out much of the chemistry in this reaction in a process that looks suspiciously like a remnant of the RNA world.

sequence specificity enter into the picture? Today the sequence specificity of protein synthesis arises because messenger RNA (mRNA) is used as a template to direct the incorporation of new amino acids into the growing polypeptide chain. Thus the nascent ribosome must also have learned how to direct synthesis of polypeptides in a template-directed fashion. Presumably this development occurred when the nascent ribosome found that it could bind its primitive tRNA substrates more tightly when they, in turn, were bound to yet another RNA, which would be the first messenger RNA. The precise sequence of this new messenger RNA would then determine the sequence of amino acids incorporated into newly synthesized polypeptide. The advent of sequence-specific polypeptide synthesis would have provided a very significant selective advantage. Namely, this advance allowed the cell to make complex, sequence-specific, and highly efficient protein-based catalysts and, because a protein's sequence is encoded in a gene, to pass this selective advantage on to its offspring. With this, we have the first true ribosome.

The Genetic Code

The ribosome translates the sequence of nucleotides in an mRNA into the corresponding sequence of amino acids that makes up a given protein. The translation table that relates RNA sequences to amino acid sequences is called the *genetic code.* The genetic code is simply the conversion table that links the sixty-four possible combinations of three bases (called triplets) present in mRNA to the twenty amino acids generally used in proteins, plus the "chain termination," or "stop," codons that tell the ribosome to halt polypeptide synthesis when the job is done (table 6.1). When it was first deciphered in the late 1960s, the genetic code seemed to give no clues as to its early evolution. Indeed, when they initially discovered that the same code applies in all of the organisms that had been studied, most researchers believed that it simply reflected a "frozen accident." That is, the thought was that the universal code that assigns a given set of base triplets to each amino acid was not fundamentally different from any of the other approximately 10^{20} possible patterns mapping the codons to the amino acids, but that once the existing code was randomly set down some 3.5 billion years ago, the cost of changing it would have been so prohibitive that it was essentially frozen. Exchanging the coding assignment of one triplet, for instance, would alter the corresponding amino acid in hundreds of proteins simultaneously, almost certainly with fatal consequences.

A quarter of a century later, however, a slightly more dynamic view of the code began to take hold. For one thing, it is now clear that a small

TABLE 6.1
The universal genetic code

Second base in triplet codon (columns); first base in triplet codon (left); third base in triplet codon (right)

First base	U	C	A	G	Third base
U	Phe	Ser	Tyr	Cys	U
U	Phe	Ser	Tyr	Cys	C
U	Leu	Ser	Stop	Stop	A
U	Leu	Ser	Stop	Trp	G
C	Leu	Pro	His	Arg	U
C	Leu	Pro	His	Arg	C
C	Leu	Pro	Gln	Arg	A
C	Leu	Pro	Gln	Arg	G
A	Ile	Thr	Asn	Ser	U
A	Ile	Thr	Asn	Ser	C
A	Ile	Thr	Lys	Arg	A
A	Met	Thr	Lys	Arg	G
G	Val	Ala	Asp	Gly	U
G	Val	Ala	Asp	Gly	C
G	Val	Ala	Glu	Gly	A
G	Val	Ala	Glu	Gly	G

number of exceptions to the code exist,* illustrating the fact that the code can evolve. Specifically, codon frequency was found to differ slightly among a few organisms, offering a viable, though rare, route to code evolution. A rare codon in a small enough genome may become so rare that it is used in only one or a few genes, and at that point it can be reassigned to a different amino acid if the change benefits the protein without producing fatal changes in other proteins. Despite these counterexamples, though, in the vast majority of cases it is effectively impossible for the codon assignments to vary on a reasonable evolutionary timescale—hence the suggestion that the current codon pattern, which presumably was originally randomly assigned, has been frozen in its current form by the nearly always fatal consequences of changing it.

But there is a problem with this "frozen accident" hypothesis. The

*These occur mainly in mitochondria, the energy-producing subcellular organelles that power our cells and contain their own DNA. In the mitochondria of starfish, for example, the codons AGA and AGG specify serine and the codon UGA specifies tryptophan, whereas in the cytoplasm they represent arginine and a stop codon. These substitutions are possible because the mitochondrial genome is quite small; it encodes only thirteen proteins. But altered genetic codes have also been found in some nuclear genomes, including some species of yeast, such as *Candida albicans*.

current genetic code seems to be far more highly optimized than one would expect were it simply an accident. That is, the current genetic code is set up such that a large fraction of mutations at the level of nucleic acid sequence are "silent," or at least chemically conservative, at the amino acid level. This is easy to see by looking at the genetic code (table 6.1); for example, look closely at each of the sixteen *four-codon blocks.* The second block of the first row, for example, represents the four codons with the sequence UCN, where "N" implies any of the four bases A, G, C or U. All four of these codons encode the amino acid serine. Thus *any* mutation of the third position in these codons will be silent! This reduces the number of possibly deleterious mutations in a serine codon by a factor of 3. And since this trait is shared by eight of the twenty amino acids and partially by several others, more than a quarter of all possible point (single-base) mutations do not alter the amino acid sequence that the DNA encodes. Moreover, many other mutations are *conservative;* mutation of the first base of many codons, for example, tends to swap chemically similar amino acids such as leucine (CUN, where, again, N denotes any of the four bases) for isoleucine (AU followed by U, C, or A) or methionine (AUG).

The robustness to mutation that is captured in the current genetic code provides a significant advantage over alternative genetic codes that lack this property. Keeping track of silent mutations and using various chemical scoring functions (like hydrophobicity) to rank the effect of potentially conservative substitutions, theoreticians have found that the current genetic code is *highly* optimized in terms of suppressing the effects of mutation. In fact, the probability of picking by chance a genetic code as error proof as ours would be close to one in a billion, which suggests that the present-day code is, somehow, the product of intensive evolutionary tweaking. But this finding seems highly paradoxical. If changes to the code are almost invariably fatal, how can the code evolve? Hints regarding the answer to this question are apparent in the structure of the code itself.

Look again at table 6.1. Notice that the amino acids in the first column, those encoded by codons with the structure NUN, are the hydrophobic (water-hating) amino acids phenylalanine, leucine, isoleucine, methionine, and valine. Similarly, all of the amino acids in the third column, those encoded by codons with the structure NAN, are hydrophilic. This has led to the suggestion that the first code was very simple; it may have used three nucleotides in a codon, but only the middle of the three *mattered.* If the middle nucleotide were a U, a hydrophobic amino acid (perhaps the simplest, valine, but we cannot know for sure) was encoded; if the middle position were an A, a hydrophilic amino acid (again, perhaps the simple amino acid aspartate, but we do not know for sure). The earliest code might thus have encoded only four different amino acids, and the earliest proteins might have been, consistent with our earlier arguments about the origins of translation itself, of simple composition.

How, then, did we go from this simple, four-amino-acid, "only the middle position counts" code to the current twenty-amino-acid code? In time, selective pressures would ensure that new amino acids became recruited into the process of building proteins. When new amino acids were added, more complexity needed to be added to the genetic code to accommodate them. According to this hypothesis, this was done using the first position in the codon. Thus valine is differentiated from leucine by the first nucleotide in their GUN and CUN codon sets; the middle position encodes "hydrophobic" and the first distinguishes which of the several hydrophobic amino acids. Using just the first two codons, we can encode $4 \times 4 = 16$ amino acids. To enlarge the set beyond 16, we have to involve the third position. We see some examples of this in the current genetic code. For example, the chemically similar (negatively charged) amino acids glutamate (Glu) and aspartate (Asp) are differentiated only by the third position in their codons. But for many codon sets, the third position is entirely redundant; all four GGN codons, for example, encode glycine (Gly). It thus seems that the genetic code *did* freeze. That is, sometime after the first two codon positions became important, but before all of the third positions became distinct, organisms became complex enough that any further changes (save the rare, limited exceptions described above) became prohibitive. The freeze occurred after only twenty amino acids had become encoded, presumably because this was the balance point between the selective pressures that pushed toward greater complexity (more diversity means a better ability to solve problems) and the increasing chance that any additional changes would prove fatal.

The near-universal spread of the standard code suggests that it was already in place in the time of LUCA. This, in turn, means that also in place were all of the different tRNAs and the enzymes required to charge them with the appropriate amino acids.* At first glance this looks like bad news for the study of early molecular evolution, but the complexity of the protein biosynthesis apparatus allows researchers to study the

*Archaea lack the synthetase enzymes that charge the glutamine and asparagine tRNAs with the "correct" amino acids. They instead charge these tRNAs with the related amino acids glutamate and aspartate, and then chemically convert them to glutamine and asparagine, respectively. It has thus been proposed that LUCA was lacking these two synthetases.

evolution of the code before that time limit. How can that be? The clue lies in the observation that there are around fifty different tRNAs (not quite as many as there are codons; some tRNAs can recognize several codons differing only in the third position, a phenomenon described as "wobble"). All tRNAs share a common, L-shaped structure and a variety of specific peculiarities such as the occurrence at specific points in the molecule of rare bases other than the four common ones (e.g., a nucleobase called pseudouracil, which differs slightly from uracil). The high degree of sequence similarity among the different tRNAs suggests that they originally evolved from a much smaller set of adapter molecules, offering researchers the opportunity to study molecular origins of the tRNA family.

The tRNA synthetases seem to be telling a similar tale about the origins and further evolution of the translational machinery. Studying tRNAs and their synthetases in detail, researchers have found that the twenty different synthetase enzymes can be divided evenly into two distinct classes. The ten members within each class share a common folding pattern (the pattern in which the polypeptide is arranged into the final three-dimensional structure of the protein), as well as common structural elements such as the chemical details of the active site where catalysis takes place and the amino acids involved in recognition of the tRNA. Thus, it seems clear that all twenty tRNA synthetases evolved from just two original RNA-binding proteins, each of which dated back to the earliest days of protein synthesis. And since LUCA seems to have contained at least eighteen tRNA synthetases (two arose later in bacteria and eukaryotes, but not in archaea), tracing back the family tree from these to the two primordial enzymes should reveal molecular details from generations that preceded her.

Researchers have not yet managed to figure out the definitive tRNA synthetase family tree in detail, but there are patterns that seem to offer clues. Each of the two classes can, in turn, be subdivided into three subclasses according to more subtle structural and functional details (table 6.2). Oddly, if one compares the equivalent subclasses, such as Ia and IIa, one finds not only exactly the same number of members on both sides, but also intriguing similarities between the amino acids and even the tRNAs charged by members of the equivalent subclasses—and this despite the fact that the two classes of enzymes do not seem to be even remotely related; their amino acid sequences and final, folded structures are utterly dissimilar.

One might put this observation down to the effects of convergent evolution, in which two very different things carrying out a common function appear similar (think fish tails and whale tails; these arose sep-

TABLE 6.2
The tRNA synthetase classes

Subclass	Class I	Class II	Subclass
Ia	Leu	Ala	IIa
	Ile	Gly	
	Val	Thr	
	Met	Ser	
	Cys	Pro	
	Arg	His	
Ib	Glu	Asp	IIb
	Gln	Asn	
	(Lys)*	Lys	
Ic	Tyr	Phe	IIc
	Trp		

*A class I synthetase specific for lysine is found only in certain species of archaea and bacteria.

All others have a class II enzyme for this specificity.

arately but look similar because they perform the same function). Thus the task of activating similar tRNAs could have produced similar-looking synthetases that are, in fact, unrelated to one another. However, researchers have made a slightly spooky observation that might point to a hidden and unprecedented relationship between the two classes. If you take the gene sequence for a class Ia synthetase and the sequence of the complementary strand (cDNA) of the gene for a class IIa synthetase (i.e., not the strand of DNA that codes for the protein, but the opposite, complementary strand), the DNA sequences for the catalytically relevant portions of the synthetases—the most highly conserved stretches around the active sites—are somewhat similar.

One speculative interpretation of these observations was proposed by Lluís Ribas de Pouplana and Paul Schimmel and goes roughly as follows. In the very early days of protein biosynthesis, *both* strands of a DNA or RNA duplex coded for some of the earliest proteins. From one such duplex evolved the tRNA synthetases, and the coupling of information on both sides of the gene was kept long enough for the two classes of synthetases to retain substantial symmetry. So far, this interpretation remains unproven and contentious, but what is certain is that the evolutionary history of protein synthesis before LUCA is recorded in the sequence diversity of tRNAs and their synthetases—we just have to learn how to read it.

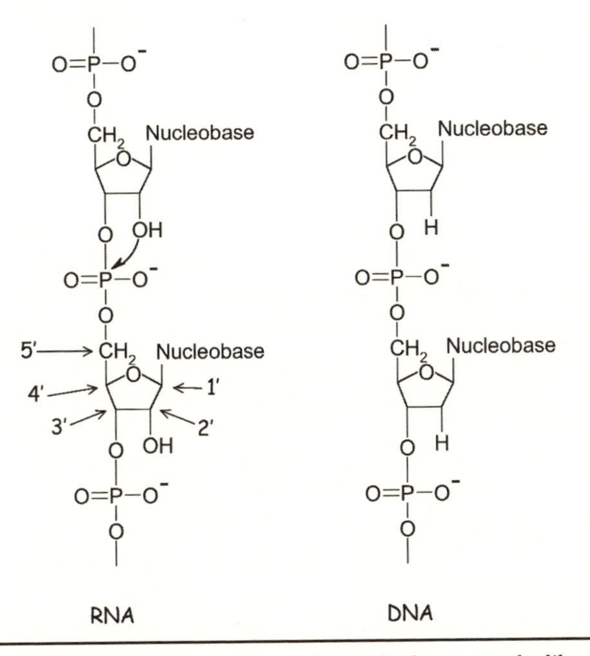

FIG. 6.2. At first glance, RNA and DNA look very much alike. But DNA is missing an oxygen atom at the 2′ position, and that makes a world of difference. As shown, the 2′-oxygen atom in ribonucleotides is perfectly placed to attack the phosphate backbone of RNA. This leads to a reasonably high rate of self-cleavage, which in turn puts a strong limit on the maximum size of an RNA-based genome. The atom numbering system, shown for ribose, holds for both RNA and DNA.

DNA Archives

In principle, RNA can serve as genetic material. In fact, it still does for many viruses, including both HIV and the West Nile virus. But the poor chemical stability of RNA limits the size of a RNA genome to a few thousand bases. Any longer, and the genome is too likely to suffer a fatal self-cleavage reaction in which the reactive, free hydroxyl group on the ribose attacks the phosphate backbone (fig. 6.2). Thus, to achieve larger genomes—perhaps to accommodate a growing number of protein-coding genes—evolution had to invent a new archival storage material.

At first glance, RNA and DNA look very much alike (fig. 6.2). What is so special about the missing oxygen atom that puts the *D* into DNA (remember: the *D* stands for "deoxy," i.e., lacking an oxygen)? Textbooks of biochemistry tend to start from DNA and then mention in passing that RNA has an —OH function in position 2′ (the prime distinguishes the atom numbering of the sugar from the numbering of the base). But it seems that, in the history of life, RNA is the earlier version of the idea, and DNA the deluxe edition introduced later.

Considering the structure of the ribose alone, removal of one of the molecule's five oxygen atoms doesn't look like such a big deal. But if you look at the ribose inside a nucleic acid polymer, you see that the 2′-oxygen is the only one that does not serve an immediate function in the primary structure. In the polymer, the oxygen in position 1′ is replaced by the nitrogen of the base, while the 4′-oxygen is holding the ring together. Oxygens 3′ and 5′ link to the phosphate groups that lead to the neighboring nucleotides. This leaves the oxygen of the 2′-hydroxyl group as the sole survivor, a chemically reactive group that might get involved in all kinds of mischief, including additional (branched) polymerization, hydrogen bonding, hydrolysis, steric hindrance, and—worst of all—the autocatalytic hydrolysis of the phosphodiester bonds we mentioned above: the 2′-oxygen is, in fact, in the perfect geometry to attack the phosphate group in the RNA backbone and cleave the backbone in two. Because of this, RNA polymers are relatively unstable and highly likely to break down under even mild conditions over the course of days or weeks. DNA lacks the 2′-hydroxyl group and thus is enormously more stable. So stable, in fact, that intact DNA has been extracted from multimillion-year-old fossils.* This enhanced stability renders DNA much better suited than RNA for the archiving of large amounts of genetic information and, because DNA is completely compatible with the RNA archives that presumably preceded it (DNA binds to a complementary RNA even better than RNA binds to its own RNA complement), it was presumably easy for this improved system of information storage to evolve from the RNA world. A vestige of DNA's takeover as the genetic material may be found in the way DNA is made in the cell: the synthesis of DNA is initiated using a short RNA "primer." The formation of RNA, in contrast, doesn't generally require a primer, and thus RNA synthesis can "bootstrap" itself.

As an additional bonus, DNA contains a new kind of nucleobase that facilitates the repair of damaged genetic material. Instead of uracil (U), DNA contains a methylated version of this base, thymine (T). The trouble with the RNA set of nucleobases is that cytosine sometimes spontaneously converts into uracil via a simple hydrolysis reaction that replaces an amino group on the nucleobase with an oxygen (see fig. 4.7). The additional methyl group in thymine allows the cell to distinguish this base from uracils that might have accidentally been created in the DNA by hy-

*Though, sadly, not yet from dinosaurs.

drolysis of cytosine. There is an entire toolkit of repair enzymes to cope with this damage: they first cut off the uracil base, then open the damaged DNA strand, and finally restore the cytosine. Thus, the introduction of thymine and the associated set of quality controls is a valuable improvement of the fidelity in genetic inheritance over the RNA world, but it would have been too costly to extend this to the "disposable" products, such as messenger RNA, which are still made of RNA today.

Which Came First, Proteins or DNA?

With the advent of a more durable genetic material and the availability of highly efficient protein-based catalysts, the machinery that copied, repaired, and transcribed the cell's genome evolved to levels of complexity that were simply not possible in an RNA world. This complexity included not only enzymes that copy DNA to make either new DNA or RNA transcripts, but also enzymes that unwind DNA, repair damage such as that induced by UV light or other sources of radiation, untangle DNA if it is all knotted up, cut it at specific sites, regulate the length of the chromosome ends, and much, much more.

But this raises a question. Given the obvious advantages that proteins and DNA possess, it is easy to rationalize why the protein-DNA world took over from the RNA world. But even if we accept that the RNA world begat the protein-DNA world, which of these new polymers came first? As yet, nobody knows whether the RNA world first recruited proteins or recruited proteins only after inventing DNA. An indirect argument in favor of DNA first is as follows: while we know LUCA used DNA, LUCA may not have contained the enzyme ribonucleotide reductase, which converts ribonucleotides into deoxyribonucleotides, because this enzyme is quite different in bacteria, archaea, and us. Steven Benner has speculated that this is because LUCA's ribonucleotide reductase was still a ribozyme. That is, it was a ribozyme that catalyzed the reaction that made DNA, not a protein; this argues that DNA came *before* proteins. Still, the ribonucleotide reductase reaction is a very difficult reaction to perform (it is one of the few biochemical reactions that relies on free radicals), so there is some question as to whether it could have been performed by a ribozyme, and no such ribozyme has been created in the laboratory. So it is probably best to say that the jury is still out on the question of which came first, DNA or protein. Or, more accurately, which came second, after RNA.

Enzymes and Metabolic Networks

At this point in our story we're talking about an organism that has in place all of the major molecular components of today's life, including

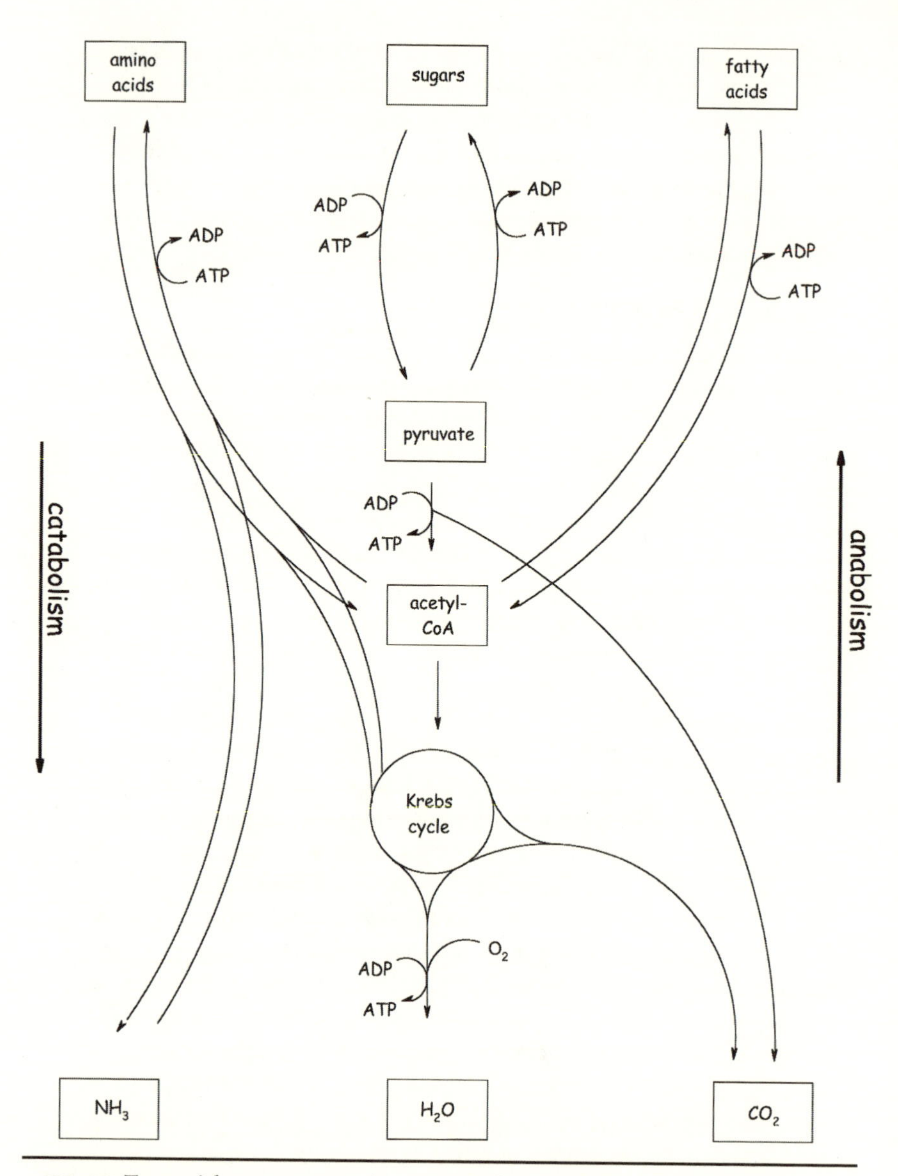

FIG. 6.3. Terrestrial genomes encode very complex metabolic networks. These include catabolic pathways, which break down large molecules (like the hexose sugar glucose) into smaller molecules (like the three-carbon pyruvate) in order to generate energy, and anabolic pathways, which synthesize amino acids, nucleobases, and the like, from simpler precursors. As an indication of the true complexity underlying this simple scheme, the Krebs cycle alone requires ten different protein catalysts.

RNA, proteins, and DNA. Given that, several natural questions then arise. How many proteins did this early organism have? What were their likely tasks? And how did they link up into complex metabolic networks? In short, of the metabolic pathways of today's cell (fig. 6.3), which would already be listed in an archaean-era biochemistry textbook?

Since the sequencing of entire bacterial genomes became feasible in 1995, genomics (the study of the full complement of genes that an organism carries) has increasingly provided researchers with new tools to address all of these questions. Obviously, the availability of unprecedented numbers of gene sequences has aided molecular phylogeny, or the construction of family trees for groups of related organisms, or even for groups of related proteins, which we discuss in the next chapter. But on a higher organizational level, the comparison of complete genomes has made it possible to investigate the question of what constitutes a minimal set of genes for a cellular organism.

The first two organisms to have their entire genomes sequenced, *Haemophilus influenzae* and *Mycoplasma genitalium*, were in part selected for these studies because their genomes are small. In fact, with just 480 genes, *M. genitalium* was assumed to contain precious little beyond the minimal set. Nonetheless, comparison of the genomes of these two organisms suggests there is a shared set of only around 260 genes that represent the minimum "essential" collection. Later studies focused on sequentially knocking out individual genes from the much larger genome (~4,100 genes) of the bacterium *Bacillus subtilis*. These studies arrived at a list of just 271 essential genes. Still other studies have produced results ranging from 150 to 670 essential genes. But it is not so much the precise number that should interest us here. Instead we should focus on the tasks that these genes fulfill and their usefulness as a model for the simplest, and perhaps earliest, metabolisms.

And what do these simple, minimal genomes tell us? In the *B. subtilis* study, most of the 271 essential genes can be clearly grouped into the broad categories of information processing (DNA processing: 27 genes; RNA processing: 14; protein synthesis: 95), cell envelope (44), cell shape and division (10), and energetics (30). Not surprisingly, this suggests that DNA replication, protein synthesis, maintenance of the cell's physical integrity, and the metabolism required for energy production are all critical elements of the simplest complete metabolism. Still, the question remains as to which of these metabolic networks we inherited from LUCA and which, if any, were invented after her. To answer this question, we have to look at the bigger picture.

Steve Benner has likened the evolution of metabolism to a palimpsest, the fancy word used in archaeology for a parchment that has been used more than once and from which traces of the imperfectly erased earlier inscriptions can still be read. For example, as we described in the previous chapter, ribonucleotide cofactors can be interpreted as vestiges of the RNA world. Benner argues that all of metabolism can be viewed this way, for, as we argued above, any metabolic pathway that is shared across all life forms was likely inherited from LUCA. Using this approach to map out LUCA's biochemistry, Benner's group found that LUCA used DNA as her genetic material and contained fairly modern-looking DNA polymerases. LUCA, as we've noted, also contained the transcriptional machinery by which messenger RNAs are made using a DNA template, and a full set of machinery for translating the messenger RNA into the appropriate protein sequences. But what about LUCA's *metabolism*? That is, beyond transcription and translation, which of the myriad of biochemical pathways that we use to convert our food into ourselves did we inherit from LUCA?

While we humans can synthesize only twelve of the twenty proteogenic amino acids (due to our rich diets, we could afford to lose some of the pathways by mutation without taking too hard a selective hit), most organisms are not similarly handicapped. Looking across the tree of life we find, in fact, not only that most organisms can synthesize all twenty amino acids, but also that the metabolic pathways by which most of the amino acids are synthesized are closely related. The only notable exceptions to this rule are the biosynthetic pathways that produce the three aromatic amino acids (phenylalanine, tyrosine, and tryptophan, each of which contains a benzene-like group in its side chain), which differ significantly between eukaryotes and bacteria. It thus seems that LUCA had the ability to synthesize most of the amino acids herself, if perhaps not the aromatic amino acids. By a similar argument, it seems that LUCA could synthesize the nucleobases; the biosynthetic routes by which both the purine and pyrimidine bases are manufactured in the cell are closely related across all three domains of life—archaea, bacteria, and eukaryotes.

But what about the mechanisms by which the cell derives its energy? We obtain most of our energy via two metabolic pathways. The first is called glycolysis and involves the *nonoxidative* breakdown of glucose into the smaller molecule pyruvate. Such nonoxidative, energy-producing reactions are called fermentation.* The ten enzymes involved in glycolysis share close relatives across the three domains of life, suggesting

*Under anaerobic conditions (such as in a champagne bottle), yeast obtain their energy from the glycolytic pathway. As a last step in the pathway, they convert the pyruvate to ethanol and carbon dioxide. Cheers!

once again that LUCA contained this key metabolic pathway. In us, the pyruvate is then oxidized to carbon dioxide in a cyclic metabolic pathway called the Krebs cycle (which we've already encountered; and more on this in the next chapter). The enzymes of our Krebs cycle, however, are not present in archaea, suggesting that LUCA did not contain this oxidative metabolic pathway. Thus it seems that LUCA performed fermentation reactions for a living.

Membranes: Wrapping It All Up

The DNA-RNA-protein division of labor is an effectively universal feature of organisms living on our planet today. Another similarly near-universal feature is the barrier that separates the living from the nonliving world: the cell membrane. Life may have originated as a set of self-propagating chemical reactions on a solid surface, in liquid droplets, or in other kinds of media (as described in chapter 5), but it acquired the ability to grow and spread independent of the medium only when it succeeded in isolating itself from the environment by creating the cell membrane. The membrane also serves the critical role of keeping the genetic material physically linked to the metabolic catalysts that it encodes. Without this linkage, the metabolic networks of the cell would not provide a selective advantage for the organism (which is defined by its genes) and evolution would grind to a halt. In a sense, the membrane was the key step in creating cellular life as we know it today. (Viruses, many of which are also surrounded by a membrane, are a later development, based on the DNA/RNA of already existing cellular organisms.)

But where did the membrane come from? The universality of the DNA-RNA-protein system strongly suggests that this system pre-dates the common ancestors of today's organisms. Similarly, all living cells are surrounded and defined by at least one double-layer lipid membrane, supporting the arguments that LUCA must have had a membrane too. The precise chemical composition of the lipids constituting the membrane, however, differs among the three domains of life. We eukaryotes and our bacterial brethren use diacylglycerides, consisting of a glycerol to which two fatty acid tails are linked via an ester bond (fig. 6.4). Archaea, in contrast, cannot synthesize fatty acids—suggesting, perhaps, that LUCA could not synthesize them—and instead build their membranes from a class of lipids called terpenoids. All three of these broadest branches of the tree of life can synthesize terpenoids, suggesting that LUCA could as well. Perhaps, then, she built her membranes as our modern archaeal cousins do. Whatever the truth about this matter, the diversity of membrane chemistry suggests that membrane optimization was still in flux when the major lineages diverged.

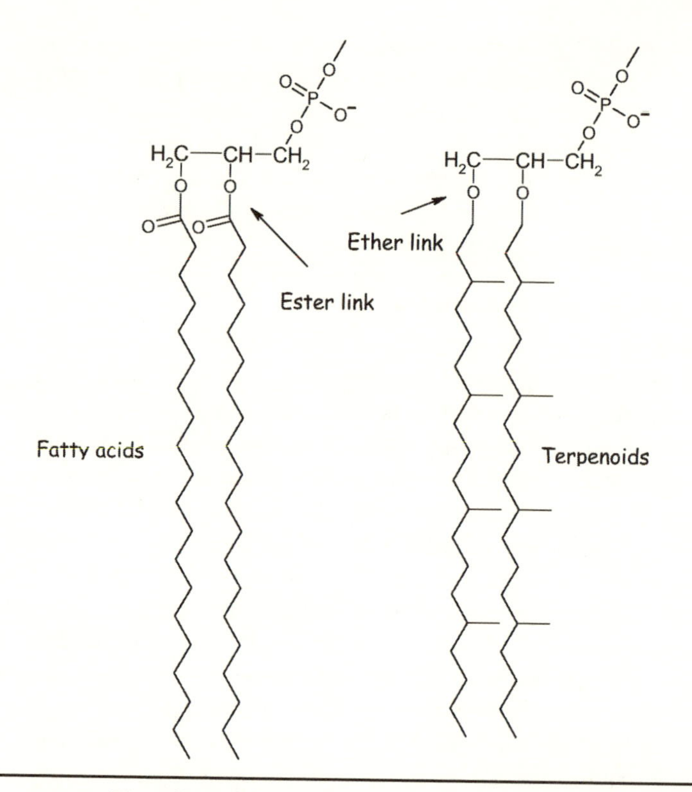

FIG. 6.4. The cell membranes of both eukaryotes and bacteria consist largely of diacylglycerides (left), in which two water-hating fatty acid tails are linked by ester bonds to a glycerol that, in turn, is connected to a water-loving "head group" such as phosphate (as shown here). In contrast, the cell membranes of archaea are formed from branched hydrocarbons called terpenoids attached to a glycerol via ether linkages (right).

Conclusions

From the first simple, self-replicating molecules to protein- and DNA-based organisms of breathtaking metabolic complexity—it's a fascinating story, the broadest sweep of which seems clear: the incessant push of selective pressure, fighting against the imperatives of chemical reactivity, guided the earliest, simplest life into the complex, robust life forms we see around us today.

As illustrated by the scientific biography of Francis Crick, who died in July 2004 at the age of eighty-eight, the tale of how we came to understand the origins of cellular life is similarly a complex and fascinating story, one that was driven by episodes of startlingly original, outside-the-box thinking. And while we may be tempted to deride those wild ideas that turned out to be off the mark, at the frontiers of our knowledge many wild ideas turn out to be true. The origins and early evolution of life is one such frontier where this kind of unbridled creativity is still useful.

Further Reading

Directed panspermia. Crick, Francis H. C., and Orgel, Leslie E. "Directed panspermia." *Icarus* 19 (1973): 341–46.

RNA evolution. Joyce, G. F. "The antiquity of RNA-based evolution." *Nature* 418 (2002): 214.

Protein biosynthesis. Nierhaus, K. H., and Wilson, D. N. (eds.). *Protein Synthesis and Ribosome Structure.* Weinheim, Germany: Wiley-VCH, 2004.

LUCA's metabolism. Benner, S. A., Ellington, A. D., and Tauer, A. "Modern metabolism as a palimpsest of the RNA world." *Proceedings of the National Academy of Sciences USA* 86 (1989): 7054–58.

CHAPTER 7

A Concise History of Life on Earth

In 1969, Carl Woese, then a young professor at the University of Illinois, came up with an unusual way to investigate phylogeny, the study of the interrelatedness of things, and in doing so uncovered a rather surprising result. For the preceding century life had been divided into four eukaryotic (cells with nuclei) kingdoms, the animals, plants, fungi, and protists (the single-celled eukaryotes), and one prokaryotic (cell lacking a nucleus) domain, the bacteria. This classification, which was based on gross cellular features, seemed only natural from our perspective as big, lumbering eukaryotes; clearly, the major divisions of life on Earth should be weighted heavily toward us and our closer brethren. Woese's ambition was to probe the relatedness of these kingdoms in more detail and to settle the question once and for all as to how the bacteria fit into the bigger picture.

Because bacteria don't have gross structural features that can easily be compared, Woese realized he would have to study phylogeny at the molecular rather than the cellular level. Even as late as the late 1960s, though, biologists had depressingly few molecules with which to perform such comparative molecular biology: at the time, the atomic-resolution structures of only about ten proteins were known,* along with the amino acid sequences of a few dozen more. Even simply establishing the sequence of a small protein took years of effort, and there were no viable methods for sequencing genes, much less entire genomes. How, then, could one compare organisms on a molecular level and work out their evolutionary relationships beyond those that are obvious from outward appearances?

To start, Woese needed a molecule that is present in all living things. As all cellular organisms have ribosomes, he chose a ribosomal RNA (rRNA): the RNA strand of the small ribosomal subunit (see fig. 6.1), known as the 16S RNA in bacteria. But instead of reading the nucleotide sequence of the molecule, which was well beyond the technology of the day, he shredded the molecule. More specifically, he digested the rRNA

*Today we are at thirty thousand and counting.

with an enzyme that cuts after the nucleotide guanosine (G), and only after guanosine. Thus the resulting fragments, short enough to be sequenced with available methods, would all constitute "words" ending with G: AUG, CG, ACACACUUG, and so on.

After sorting the fragments by their length (using gel electrophoresis, a well-established method of separating molecules according to their sizes), sequencing some of them, and thinking about the results, Woese found that the most useful words for his purpose were those of six letters or more—shorter words were too common to provide clues as to who was related to whom. There are $3^5 = 243$ different six-letter words ending in G, and a typical 16S RNA contains around 25 of them. These short words thus provided good enough statistics to allow Woese to "fingerprint" organisms by their 16S RNA and to use these fingerprints to map out the relationships among organisms.

Over many years, Woese and his graduate students compiled "dictionaries" of 16S RNA words from different bacterial species and constructed family trees based on the degree of identity between the content of the dictionaries. To their great surprise, a subset of the bacteria they studied, namely those that produce methane, turned out to be just as distantly related to other bacteria as they are to elephants, guinea pigs, or ourselves. Woese proposed that these microbes represented a new branch on the tree of life, equal in stature to the Eukarya and Bacteria, which he called the Archaebacteria (later shortened to Archaea) because it seemed to represent an "ancient" form of life.

But did the world of microbiology accept his new World View? Of course not. After all, Woese's new-fangled "molecular fingerprinting" aside, Archaea and Bacteria look very similar from the perspective of us big, multicellular organisms, and thus, naturally, Woese's arguments were rejected more or less out of hand. It was not until the late 1990s, when the first archaeal genomes were sequenced, that Woese's claim that Archaea were a separate "kingdom," and that life is better divided into two prokaryotic and one eukaryotic "domains" rather than four eukaryotic and one prokaryotic "kingdoms," finally took hold. With literally thousands of gene sequences to study, comparative molecular studies demonstrated quite compellingly that Woese was correct, and the Archaea are as distant from Bacteria as they are from us. The wheels of science sometimes turn slowly, but turn they do.

How Old Is Life on Earth?

On the one hand, the history of life on Earth is a somewhat parochial topic for wide-ranging types like astrobiologists. Clearly, for example, the division of Terrestrial life into three domains would hardly be of relevance to the study of, say, life on Jupiter's moon Europa (if there is any!). On the other hand, the history of how the story unfolded here is the only example we have, so if we're careful not to be complacent, not to fall into thinking that how it occurred on Earth is the only way it could have occurred here or anywhere else, the topic seems worthy of serious consideration.

And how and when did the history of life on Earth unfold? Although the Earth formed some 4.56 billion years ago, and its crust probably solidified some 100 million years later, after the formation of the Moon (chapter 3), it was effectively uninhabitable for long after that. It was not until some 3.8 billion years ago that the late heavy bombardment and its planet-sterilizing impacts came to an end. A key question, then, is: how quickly after the end of the late heavy bombardment did life arise on our planet?

The answer to that question is obscured, at least in part, by our lack of detailed knowledge of the Earth's early history. Even after the end of its turbulent formative years, the Earth remained (and remains) a highly dynamic planet, and thus few records of its early days have survived intact. Due to the incessant erosion brought on by the Earth's hydrological cycle, and plate tectonics constantly recycling crust into the mantle, typical rocks on the surface of our planet are estimated to have a half-life of only a few hundred million years. The chances of the Earth's first rocks—those dating from the so-called Hadean Era (table 7.1, a timeline of the history of life on Earth, lists the geological intervals)—surviving this gauntlet and remaining unscathed to the present seem to be nil, because no significant rocks older than 4 billion years have been identified. It is only from the Archaean Era, which started about 3.6 billion years ago, that much of a geological record has been preserved.

The oldest recognized rocks on Earth are found in two locations in North America. The older of the two are the Acasta gneisses (metamorphosed igneous rocks; i.e., rocks that originated from molten material and were later altered) from near Great Slave Lake in northern Canada. Unfortunately, though, during the 4.03 billion years since these rocks solidified they have been modified by heat and pressure to such an extent that they provide little information about what the Earth was like that early in its youth. More critical to our story are the slightly younger, apparently supracrustal rocks of Isua and Akilia, West Greenland. The *supracrustal* tag denotes that these metamorphosed rocks were deposited as sediments or volcanic flows in shallow water when they were formed, some 3.7–3.8 billion years ago. If this mineralogical assignment is correct, these rocks confirm that liquid water (viewed, of course, as

TABLE 7.1

A timeline of the history of life on Earth

Geological era/period			Time (years ago)	What was up?
Hadean			4.56 billion	Formation of Earth
			~4.2 billion	Formation of Moon
			4.05–3.70 billion	Formation of oldest rocks still in existence
			3.8 billion	End of late heavy bombardment
Archaean			3.6 billion	Formation of continents
			3.5 billion	Formation of first putative microfossils/ stromatolites
			2.2–2.4 billion	First hints of oxygen
Proterozoic			~2.7 billion	Formation of Australian oil shales; molecular fossils of first eukaryotes?
			2.0–2.2 billion	Oxygen levels climb to ~18%
			1.7–1.9 billion	Putative first eukaryotic fossils
			~1.2 billion	Invention of sex
			1.2 billion	First multicellular organisms
			960 million	Divergence of plants, animals, fungi
Phanerozoic	Paleozoic	Cambrian	542 million	Cambrian "explosions"
		Ordovician	488 million	Trilobites rule the ocean
		Silurian	443 million	Invasion of the land
		Devonian	405 million	Fishes diversify Origins of amphibians
		Carboniferous	360 million	Forests of tree ferns
		Permian	290 million	Period ends with largest recorded mass extinction
	Mesozoic	Triassic	251 million	Origin of dinosaurs
		Jurassic	215 million	Age of dinosaurs
		Cretaceous	135 million	Age of dinosaurs
	Cenozoic	Tertiary	65 million	Dinosaurs wiped out
			7 million	Human/chimp divergence
		Quaternary	2 million	*Homo erectus*
			0 million	You're reading this

critical for the formation of life) existed on Earth at that time. But was there life in this water?

Sediments are formed from a steady rain of material—both organic and inorganic—that falls from the water, so they provide an ideal place to look for signs of past life. Perhaps consistent with this, the Isua and Akilia rocks contain small globules of graphite, the pure carbon form used in pencil "lead." Is this evidence that life was flourishing more than 3.7 billion years ago? As it contains only carbon and none of the other chemical elements necessary for life, graphite is not usually associated with biology. On the other hand, this particular graphite might be. The reason is that the Isua and Akilia rocks have been significantly metamorphosed; had they originally contained life, the organic carbon would have been dehydrogenated to form graphite when the rocks became buried and "cooked" deep within the Earth. Based on this argument, the German geologist Manfred Schidlowski, and later the American Stephen Mojzsis, suggested that the carbon extracted from these rocks might have been derived from living things. Mojzsis, at the time a graduate geochemistry student at the Scripps Institution of Oceanography, working under Gustaf Arrhenius (grandson of the Arrhenius of panspermia fame; see chapter 5), characterized the ratio of the carbon isotopes ^{12}C and ^{13}C in these ancient materials by heating the rock and analyzing the carbon compounds that were driven off. What they found was carbon depleted in the heavier isotope *as is observed today in the organic carbon compounds produced by photosynthesis*! (The more rapidly moving, lighter carbon isotope is preferentially reduced in the photosynthetic reaction.) Thus, they suggested, not only had life arisen at the time the original rock was deposited—just a few hundred million years after the crust cooled—but it had evolved to such a high degree of complexity that photosynthesis was already an important and common form of metabolism.

But are these putative indications of life, much less photosynthesis, on firm footing? Within only a few years of the publication of Mojzsis's investigations, other researchers began to question the evidence on numerous grounds. For example, the seemingly telling graphite occurs in veins of carbonate rock that were probably formed via the injection of hot fluids when the older host rocks were buried deep within the Earth, perhaps long after they were initially formed. Moreover, the isotopically odd carbon observed in the laboratory was released at a temperature far too low to be from the graphite (which has the highest vapor point of any element) and thus could well be a more recent contaminant. Even the age of the relevant rocks has been questioned—in fact, by members of Mojzsis's original research team (see sidebar 7.1)—as has the original identification of the Isua and Akilia rocks as sedimentary (sediments are a great place to collect fossils, igneous rocks are not). At best, then, the jury is very much still out regarding the evidence for life on our planet more than 3.8 billion years ago.

If the evidence for life at 3.8 billion years ago is poor, how much more recently do we have to go before the evidence becomes firmer? Perhaps not much. In 1993 William Schopf, a professor of paleobiology at the University of California, Los Angeles, described 3.46-billion-year-old specimens from Western Australia (near the ironically named, fiercely hot town of North Pole) that seemed to contain microscopic, tar-colored fossils of bacteria. The tiny organisms were encased in chert, an extremely fine-grained rock that can preserve the smallest of details. Schopf sorted the bacteria into eleven taxa, or distinct groupings, based on the shapes of the fossils, and claimed that, in terms of these shapes (called "morphology" by the paleontologists in the crowd), seven seemed to be early relatives of cyanobacteria (a type of photosynthetic bacteria). Raman spectroscopy of the samples, which crudely identifies molecular components, indicated that the tarlike substance contained within the fossils was kerogen, a complex mixture of hydrocarbons that is typically produced when biological material is subjected to heat and pressure beneath the Earth's surface. Taken together, Schopf claimed, this provides incontrovertible evidence that complex ecosystems, likely comprising multiple species of photosynthetic cyanobacteria, existed as little as about 300 million years after the end of the late heavy bombardment.

Following up on Schopf's claims, Donald Canfield and coworkers, at Odense University in Denmark, have studied sulfur isotopic fractionation in the same rocks and have found possible signatures of past life in the sulfur-containing mineral barite. If their identification proves correct, it would not only confirm the existence of life at 3.5 billion years ago but also identify one of its key metabolic reactions: namely, the use of sulfate (SO_4^{2-}) to oxidize hydrogen or hydrocarbons to produce sulfide (S^{2-}) and water or carbon dioxide.

Schopf's claims, however, have also found critics. One criticism claims the shapes of the putative fossils are ambiguous; of the thousands of inclusions in the rock, only a tiny fraction look like cyanobacteria or, indeed, any contemporary bacteria. Of course, cyanobacteria do not fossilize well, and after sitting around for 3.5 billion years, many of these might be expected to have become degraded. The second criticism was raised by Martin Brasier of Oxford University, who says Schopf misun-

SIDEBAR 7.1

The Dating Game

A small part of the debate on the ancient rocks of Greenland concerns their age. Are they really 3.8 billion years old? In chapter 3 we discussed isotopic dating, but while this technique is straightforward and well established, applying and interpreting it sometimes is not. A big part of the problem is that this method dates the last crystallization of the rock. The formation of sedimentary rock—the kind that's likely to contain fossils, and the kind putatively observed on Akilia—does not involve melting and crystallization, and thus sedimentary rock cannot be dated directly. Sometimes, however, it is possible to define the "minimum age" of sedimentary rocks by dating igneous "intrusions," veins of minerals solidified from a molten state, cutting through the sedimentary rock. Such igneous intrusions occur on Akilia and must have formed after the sedimentary rock in order to have cut through it. These same inclusions contain crystals of zirconium silicate ($ZiSiO_4$), called zircons, which can be accurately dated.

Scientists interested in dating rocks, grandly called geochronologists, often rely on zircons. This is because, when zircons crystallize from magma, they contain uranium, which is similar in size to zirconium and thus fits into the crystal lattice, but no lead, which is preferentially excluded from the crystal lattice. Why is this important? As noted in sidebar 3.1, uranium decays into lead at a known rate (or rates, actually), and because all the lead trapped in the zircon crystal lattice originally must have come from uranium, the ratio of uranium to lead reflects the time since the

derstood the geology of the supposed microfossils, which were preserved not in marine sediments, which would have collected fossils, but rather in a hydrothermal vent or even in volcanic glass, in which fossils are much less likely to form. Once again, it seems the jury is still out, although the case in favor of life's remnants in these rocks seems significantly better established than the case for earlier life.

If the evidence for life at 3.5 billion years is also contested, when does the evidence for life on Earth become incontrovertible? That's not such an easy question to answer. As we move forward in the geological record, we simply see more and more of the same for quite some time. That is, we see more and more of what look like microfossils (perhaps not so much because the putative organisms had become more plentiful but rather because the rock record itself becomes more plentiful) as we move from 3.5 billion years ago toward the present. For example, 3.4-billion-year-old rocks from Africa preserve many bacteria-sized spheres, some of which seem to have been caught in the process of dividing. And in early 2000, the Australian geologist Birger Rasmussen, now at the Mas-

zircons formed. More specifically, because ^{238}U decays into ^{206}Pb with a half-life of 4.47 billion years, and ^{235}U decays into ^{207}Pb with a half-life of 0.7 billion years, these uranium-lead "clocks" provide two independent measurements of a zircon's age. Thomas Krogh, who helped develop the uranium-lead zircon dating method at the Royal Ontario Museum in Toronto, says that, even if the zircons are reheated, as happened at least once to the Greenland samples, they retain a "memory" of their first crystallization.

Obviously, though, dating igneous zircons can set only the minimum age of a sedimentary rock, and even then only if the relationship between the igneous intrusion and the sedimentary substrate is well understood. But because the Greenland rocks were severely deformed during their nearly 4 billion years on Earth, the sequence of their formation has become jumbled. Considering this, Stephen Moorbath, a geologist at Oxford University, contends that the sedimentary rocks were most likely deposited "only" 3.65–3.70 billion years ago. This slightly more recent dating would explain the absence of the element iridium—rare on Earth but common in asteroids—or any other signs of the late heavy bombardment that would have been expected if the rocks were, as initially thought, more than 3.8 billion years old.

On a separate note, geochronologists working on Australian sediments have found a single, small zircon crystal that apparently withstood the erosion that created the original sediments and thus pre-dates them. This micrometer-sized crystal has been dated at 4.4 billion years and, albeit small, is the oldest known Terrestrial "rock."

Moorbath, Stephen. "Palaeobiology: dating earliest life." *Nature* 434 (2005): 155.

sachusetts Institute of Technology, reported convincingly lifelike microfilaments in some 3.2-billion-year-old Australian sediments. But it is not until 2.7 billion years ago (more than a billion years after the end of the late heavy bombardment) that truly compelling evidence was laid down, again in what is now Australia.* Roger Summons of the Australian Geological Society and Roger Buick, now at the University of Washington, found oil shales from this period that are exceptionally well preserved and unusually rich in organic matter. And as we will see below, this organic material contains what may be the oldest unambiguous signatures of life on Earth.

Evidence in favor of life on Earth rapidly increases after 2.7 billion years ago. The supporting evidence cropping up at this time includes some "molecular fossils," including many that record the formation of free oxygen in our atmosphere. As we discussed in chapter 3, abiologi-

*Australia has had relatively quiescent geology and thus has the best-preserved rock record. It also lacks much in the way of mountains, for the same reason.

cal processes such as the photolysis of water tend to oxidize a planet's atmosphere over geological time. But it takes life—more specifically, photosynthesis—to push an atmosphere all the way over into oxic. Thus, by dating that transition we can at least set a lower limit on the advent of photosynthesis.

The geological record provides a history of free oxygen on Earth. An important part of this record is contained in paleosols, ancient soils that have been turned into rock. Paleosols laid down before 2.7 billion years ago contain significant amounts of the iron mineral pyrite (FeS_2) and the uranium mineral uraninite (UO_2), and since these minerals oxidize rapidly in the presence of O_2, we know they must have formed under anoxic conditions. James Farquhar of the University of Maryland has noted that the isotopic ratios in sulfates found in rocks also provide a clue to the free oxygen content of the Earth's early atmosphere: photochemical reactions can shuffle the isotopic composition of atmospheric sulfur dioxide in a characteristic manner, but free oxygen would destroy the pattern before the sulfur could make it down to the planet's surface to become locked into rocks. Based on this, Farquhar has argued that the free oxygen concentration in the atmosphere could not have risen above a paltry one part per million before some 2.4 billion years ago.

Lastly, starting more than 3 billion years ago and lasting for at least a billion years, we had banded iron formations (BIFs), which, incidentally, are the dominant commercial iron ore. These enormous formations consist of alternating layers of deep red ferric oxide (rust) and the silicate mineral chert. BIFs form when highly soluble ferrous iron (Fe^{2+}) is oxidized to ferric iron (Fe^{3+}), which in turn forms an insoluble precipitate (which is why you cannot wash rust off your car). Current thinking is that BIFs represent the global transport of iron in the ancient ocean to sites at which oxygen was being produced (perhaps by photosynthesis, perhaps by the photolysis of water in the atmosphere, or perhaps by both), oxidizing the iron, causing it to fall out of solution. After a billion years, though, the ocean's iron became depleted and the formation of BIFs stopped. This presumably corresponded with the advent of an oxic atmosphere and, indeed, red, ferric-iron-containing sediments are a common feature of the geological record from about 2 billion years ago to the present. Thus it seems that, by a couple of billion years ago, photosynthesis was so common that it had begun to dominate even the planet's geology.

The First Complex Ecosystems

Before the ozone layer came into existence, bacterial life could exist only underground or under the cover of at least a few centimeters of water, a

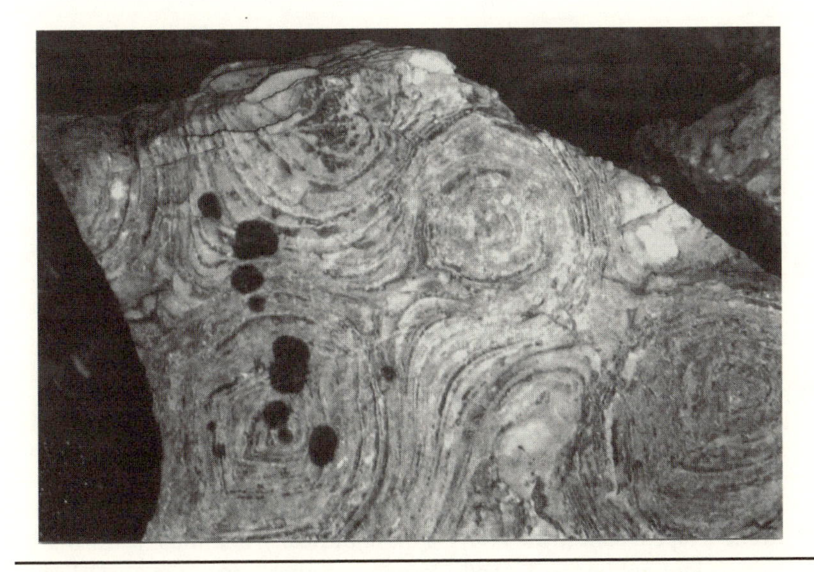

FIG. 7.1. The tip of the author's boot (on the left) on top of some billion-year-old stromatolite fossils in Glacier National Park, in America's Rocky Mountains.

restriction that somewhat limits the success that can be achieved even with a photosynthetic lifestyle. Perhaps in response to this, the cyanobacteria seem to have invented the first organized supracellular structures, and with them the oldest truly compelling fossils: stromatolites. Stromatolites are meter-tall dome-shaped or conical formations of finely layered sedimentary rock, often forming concentric structures (fig. 7.1). They are generally believed to have been formed by mats of photosynthetic bacteria such as cyanobacteria. However, if these structures are to serve as unambiguous indicators of early life, the presence of typical fine structures or microfossils is often considered essential, as there are some indications that similar sedimentation patterns may have arisen abiotically as well.

While stromatolite fossils are a dominant feature of many Precambrian sedimentary rocks, living examples are relatively rare today. With the rise of animals, growth as a thick, delicate, defenseless (and tasty?) mat of bacterial cells is not as clever a lifestyle as it once was. Thus, today, stromatolites are limited to a few select niches, typically in waters that are too saline to allow grazing animals to eat or otherwise disrupt them. For example, Hamelin's Pool, at Shark Bay on the coast of Western Australia, is a highly saline marine environment in which stromatolites still thrive, providing an opportunity to see how ancient stromatolites might have formed. There, communities of microorganisms—usually cyanobacteria but sometimes eukaryotic algae as well—spread out in coherent mats across the surface of sediments or rocks. The cells produce a thick, mucus-like material that glues them all together and affixes them to the surface. The mucus also traps fine sediments carried in the waves and currents. As this layer of sediment accumulates, the cells grow or migrate upward in order to continue photosynthesis. Cells remaining behind are cut off from the light and die. Other organisms consume the organic material from the dying cyanobacteria, in turn liberating carbon dioxide. The carbon dioxide reacts with water to form carbonic acid, which binds calcium and precipitates out as a layer of limestone.

The oldest putative stromatolite fossils are found in the 3.5-billion-year-old rocks near North Pole, Australia, the hotspot we mentioned above. But the interpretation of these finely layered rock formations as being biological in origin remains at least a bit contentious. After all, abiological processes produce finely layered rocks too (albeit making dome-shaped, concentric layers is more difficult), and thus the formations might not be fossils at all. Compounding the issue, the Australian stromatolites lack any clear indication of microscopic fossils. In contrast, stromatolites from the early Proterozoic, dating from about 2.5 to 1.6 billion years ago, are much more common and sometimes contain fairly convincing evidence of microfossils. After this time, stromatolites continued to dominate the fossil record until just before the end of the Proterozoic (i.e., just before the Cambrian explosion), some 600 million years ago, when animals proliferated and presumably began to graze on them or, at the very least, disturb them by crawling over them and disrupting their fragile organization.

When Did LUCA Live?

The paleontological record is not the only record we have of the history of life on Earth. By identifying metabolic pathways that are held in common across all life, we earlier defined the likely metabolic "toolkit" of LUCA, the last common ancestor of all life on Earth (chapter 6). Inspection of what LUCA's metabolism did, and did not, include allows us to hazard a guess as to when she lived. For example, a number of LUCA's metabolic pathways involve iron-containing enzymes that, on careful reflection, might seem like unfortunate choices to a contemporary biochemist. The problem is that in today's oxic environment iron is quickly oxidized to the ferric state (Fe^{3+}), which as described above is *extremely* insoluble. Because of this, iron is the limiting nutrient in the ocean, and marine microorganisms have had to invent an impressive arsenal of "chemical warfare agents" (called siderophores) with which to steal iron out of the grasp of other bacteria. If iron is so hard to get that its use represents a selective disadvantage, why did LUCA use it? The thought is that LUCA lived before the advent of an oxic environment, when soluble iron was plentiful. In contrast, copper in its most oxidized form (Cu^{2+}) is much more soluble than reduced copper and thus, while many more recently invented branches of metabolism employ copper-containing enzymes, LUCA seems to have avoided that element. From these and similar arguments it seems clear that LUCA pre-dated the formation of our oxic atmosphere at 2 billion years ago. But there is a big gap between that date and the first potentially solid evidence for life on Earth at 3.5 billion years ago. So, while we have bounded the problem we do not know when in this vast span of time LUCA lived.

How Photosynthesis Changed the World

Sunlight interacts with living organisms in a wide variety of ways. On the simplest level, the sunlight that hits the day-side of our planet delivers an energy flow of more than 170,000 terawatts (trillion watts), corresponding to the electricity output of 200 million nuclear power stations. Even though nearly a third of this energy is reflected straight back into space, the other two-thirds stays with us and keeps us warm in our cold Universe. It is the biggest contribution to our planetary energy balance by more than three orders of magnitude (followed by geological heating, human activity, and tidal friction). All life forms, including the earliest and most primitive ones, must have benefited from this heat supply, as it raises the surface temperature of our planet into the range that allowed water to remain liquid and life to evolve.

At the most sophisticated level, light reflected from objects around us allows us to perceive our environment with our eyes. Although a relatively recent development compared with the evolutionary timescale dominated by bacteria, vision is a reasonably straightforward ability for complex animals to come up with. So straightforward, in fact, that evolution has invented it many times, independently, with different designs.*

In between these two uses of light—one very general, the other highly specific—evolution developed a third, equally important way of making use of light, thereby starting a revolution that arguably changed the nature of life on Earth more than any other single event in its history. At some point, a group of bacteria, probably most closely related to today's cyanobacteria, came up with the two-step photosynthetic

*Octopus eyes, for example, look remarkably like our own, despite the fact that the last common ancestor we share with our excessively armed friends is thought to have been sightless.

method that not only uses the energy of light more specifically than was possible before, but also creates oxygen in the process. This is not only the oxygen that we breathe, but equally importantly the oxygen that eventually produced the stratospheric ozone layer that protects us from the Sun's hard ultraviolet and allowed multicellular organisms to finally conquer dry land.

The selective pressures in favor of photosynthesis are clear: quite simply, as LUCA and her offspring consumed the various reduced materials that were available in the environment, eventually a new source of energy had to be found. Less clear is how the photosynthetic apparatus in modern plants and algae might have arisen in the first place. As any student who has tried to memorize it will remember, the standard photosynthetic apparatus is incredibly complex. Fortunately, however, some of the simpler (and presumably older) versions of bacterial photosynthesis are still around today and can help us understand how life came up with this new technology.

The simplest and best understood light-harvesting system is that of the extremely halophilic (salt-loving) archaea from the genus *Halobacterium.* The membranes of these cells contain characteristically colored patches known as the purple membrane. Its main component is a remarkably robust protein, bacteriorhodopsin, which uses light to pump protons out of the cell. With this activity it creates a proton gradient across the membrane, which is a primitive means of storing the energy in an electrochemical form. Another membrane protein, now called (for historical reasons) the F_1F_oATPase, uses this proton gradient to drive the synthesis of the energy currency ATP. This mechanism constitutes a relatively inefficient, "hand-to-mouth" use of solar energy, as it creates only chemical fuel, not permanent chemical bonds, so it doesn't count as full-fledged photosynthesis.

In contrast, five major groups of bacteria have learned how to use the energy in sunlight to synthesize reduced carbon-containing molecules. Of note, four of these do so without producing oxygen in the process. These are the Chlorobiaceae (green sulfur bacteria, e.g., *Chlorobium*), the Thiorhodaceae (purple sulfur bacteria, e.g., *Chromatium*), the Chloroflexaceae (green nonsulfur bacteria, e.g., *Chloroflexus*), and the Athiorhodaceae (purple nonsulfur bacteria, including *Rhodopseudomonas viridis,* which provided the first ever atomic-resolution structure of a photosynthetic reaction center). Like green plants, these non-oxygen-producing photosynthetic bacteria run a redox reaction that reduces carbon dioxide (carbon at oxidation state +4) to carbohydrates (carbon at oxidation state 0). This requires a reducing agent that, in turn, gets oxidized in the reaction. In plants, this reactant is the oxygen of water (oxidation state −2), which is oxidized to molecular oxygen (oxidation state 0). By contrast, the sulfur bacteria use sulfides as reducing agents (leaving elemental sulfur as waste), and the nonsulfur photosynthetic bacteria employ hydrogen and small, reduced carbon compounds. These materials are so easily oxidized that the energy of a single photon is sufficient to extract electrons from them, and this relatively simple photosynthetic machinery (called a "photosystem") works nicely. Better still, sulfides, such as hydrogen sulfide, and other reducing materials must have been abundant in the relatively reduced environments available on the young planet.

But eventually the sulfides and hydrogen would have run out and the earliest photosynthetic organisms would have found themselves in need of a new source of electrons to run their reactions. A potentially obvious source is reduced carbon compounds. Indeed, as we just mentioned, even today some bacteria use readily oxidizable organic molecules, such as isopropanol, as electron donors, which they convert into more oxidized organics such as acetone. But these donors have to be found or made first, and if the oxidized waste product is not useful to the cell, the entire process becomes uneconomical (you waste a molecule containing several carbon atoms just to catch one carbon from CO_2). A more promising—but also more ambitious—solution would be to use water as the reductant, as its relatively unfavorable oxidation potential is balanced by its extremely favorable abundance. There was just the one hurdle to overcome: the oxygen in water holds its electrons so tightly that the energy in *two* photons of visible light is required to wrest them free. Cyanobacteria—the fifth, final, and most successful group of photosynthetic microbes—came up with the solution to this by hooking up two photosystems (PSI, PSII), each of which can trap one photon and convert its energy into chemical energy, in a serial arrangement (a system inherited by plants; see fig. 7.2). This allows the energy in two photons to be summed in order to achieve the oxidation of water to oxygen and the efficient exploitation of the liberated electrons and protons.

The resulting "two-stroke," light-driven "engine" is one of the most complicated molecular machines we know. Essentially, the light energy captured by the chlorophyll molecule of photosystem II (PSII) lifts an electron to a higher energy level. In a slower reaction, the resulting "hole" is filled with an electron pulled from water (ultimately, after four electrons have been sequentially removed from two water molecules, releasing an oxygen molecule). The high-energy electron pair flows down a cascade of reactions with the overall effect that it is ultimately transferred to photosystem I (PSI), producing in net one molecule of ATP in

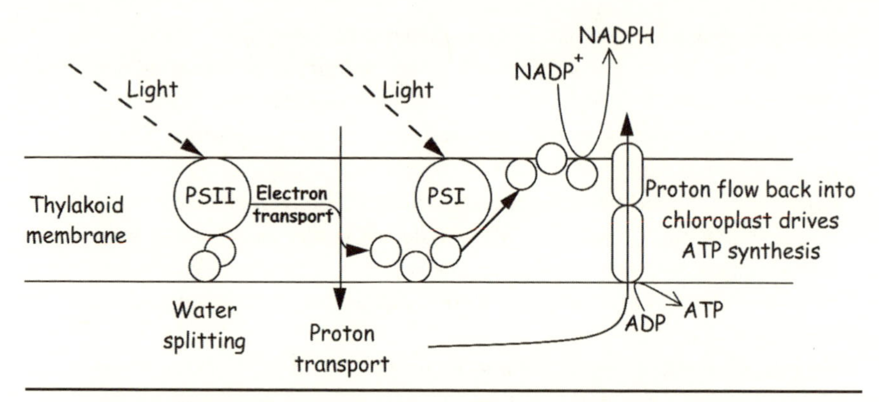

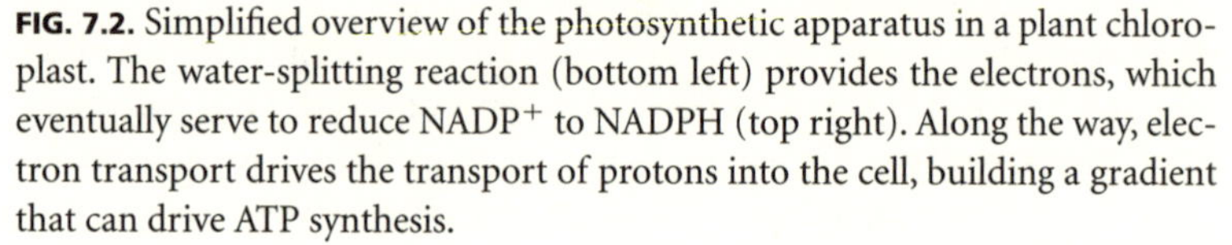
FIG. 7.2. Simplified overview of the photosynthetic apparatus in a plant chloroplast. The water-splitting reaction (bottom left) provides the electrons, which eventually serve to reduce $NADP^+$ to NADPH (top right). Along the way, electron transport drives the transport of protons into the cell, building a gradient that can drive ATP synthesis.

the process. And while some of the electron's new-found energy is given up (ultimately to form the ATP), it still arrives at PSI in a more energetic state than at its starting point. In the second light reaction, the active center of PSI absorbs a photon and lifts this now higher-energy electron to a still higher energy state. The electron, combined with a proton derived from the split water molecule, then reduces the redox carrier $NADP^+$ to form NADPH, the ribonucleotide that stores reduction potential in all cells.

The NADPH produced by the light reactions carries the reduction potential produced by photosynthesis over to the "dark reactions" (called this because they do not directly require the input of light), where it is used in the Calvin cycle. This biochemical cycle—named after Melvin Calvin (1911–97) of the University of California, Berkeley, who in 1961 won the Nobel Prize in Chemistry for its discovery*—"fixes" the carbon dioxide by covalently attaching it to the sugar ribulose-1,5-bisphosphate, splitting it into two 3-phosphoglycerate molecules. The 3-phophoglycerate is, in turn, reduced to form glyceraldehyde-3-phosphate, from which the fixed carbon now enters a complex metabolic shuffle between ten sugar intermediates in all. Ultimately, for every six carbon dioxide molecules that enter the pathway, two glyceraldehyde-3-

*Calvin elucidated the so-called dark reactions of photosynthesis using ^{14}C to trace the path of carbon from CO_2 to the final sugar. This study, one of the first uses of radioactive tracers in biochemistry, was greatly aided by the fact that ^{14}C was first synthesized at the Berkeley Synchrotron just a few years earlier.

phospate molecules are split off (to serve as the starting material for the synthesis of other sugars, such as the hexoses, amino acids, nucleobases, and all the other carbon-containing molecules the organism needs) and one molecule of ribulose-1,5-bisphosphate is regenerated. The latter allows the cycle to start anew. Overall, the formation of one six-carbon hexose, such as glucose, requires 12 molecules of NADPH, which in turn were synthesized from 12 electron pairs (generated using 48 photons) from the light reactions.

Comparison of plant photosystems with those of the more primitive bacteria mentioned above shows that those of the green bacteria resemble PSI, while those of the purple bacteria resemble PSII. Studies of the similarities and differences between the various photosynthetic systems suggest an evolutionary history: photosystems I and II arose in different branches of the bacterial family tree. The ancestors of modern cyanobacteria first had just one of these systems and only later acquired the second by horizontal gene transfer (cyanobacteria are notoriously efficient at pirating genes). After acquiring the second, they managed to couple the two in a manner that enabled them to use water as the reductant.

The advent of our oxic atmosphere after the invention of photosynthesis had wide-ranging consequences for the anaerobic bacteria that until then had dominated the biosphere. Indeed, the advent of an oxygen atmosphere has been called the greatest environmental catastrophe in the history of our planet, a catastrophe that killed off many branches of the tree of life and pruned many others back to the few remaining anaerobic niches (in anoxic muds, for example). The change was sufficiently drawn out over time, however, that even 700 million years later the oxygen concentration was still less than one-fifth of what it is today. The gradual increase in atmospheric oxygen provided sufficient time for some life to adapt to oxic conditions, and indeed to benefit from the new opportunities it created.

One important consequence of the oxygen revolution is the evolution of a sophisticated "double cycle" of life, in which the early cyanobacteria produced carbohydrates and oxygen from light and carbon dioxide while the ancestors of mitochondria learned to burn the carbohydrates using oxygen (in the reactions of the Krebs cycle, which we discuss below), producing the carbon dioxide that the photosynthetic organisms needed. This fundamental double cycle provided the metabolic basis for the evolution of multicellular organisms. Without the oxygen revolution, metabolism would be limited to relatively low-energy fermentation reactions and life would probably have remained limited to simple, single-celled organisms.

Yet another important opportunity created by the oxygen revolution was the chance to colonize land. The only reason living organisms can thrive on dry land without being fried by the high-energy parts of the solar spectrum is the ozone content of the stratosphere. Although the phenomenon is often referred to as "the ozone layer" and even measured in terms of the thickness it would have if there were such a thing, the ozone is in fact rather dilute, and the compound as such is highly unstable. But its fleeting presence in the stratosphere (which extends between the heights of 16 and 50 km) is sufficient to absorb the most damaging parts of the far ultraviolet solar radiation.

The Advent of Aerobic Metabolism

The ancestors of cyanobacteria learned how to turn energy (from sunlight), carbon dioxide, and water into carbohydrates and oxygen. In the process, they created ecological niches for other organisms running the reverse reaction: burning carbohydrates to produce energy. Oxygen-producing photosynthesis paved the way for oxygen-consuming metabolism.

Remarkably, each of these fundamental processes revolves around a circular biochemical pathway in which a small organic molecule acts as a matrix to which carbon is added, only later to be removed in some other guise. In photosynthesis, the foundation molecule is the five-carbon sugar ribulose, to which carbon dioxide is added to form two compounds with three carbons each, which ultimately feed the production of six-carbon sugars (fructose) and the restoration of the ribulose carrier. Oxidative digestion uses the four-carbon compound oxaloacetate as a matrix, which reacts with two reduced carbons in the form of an activated acetic acid to form citric acid, which gives the cycle one of its names. It is also known as the TCA (tricarboxylic acid) cycle, and as the Krebs cycle after Hans Krebs (1900–81), who discovered it in 1937, following up on his earlier discovery of the urea cycle.* The citric acid is then oxidized in a series of ten steps, producing two molecules of carbon dioxide and, ultimately, another molecule of oxaloacetate to continue the cycle.

In animals, the Krebs cycle is localized in the mitochondria. It is "fed" with acetyl-CoA, an activated form of the two-carbon molecule acetic acid, by the pathways that degrade sugars (glycolysis) and fatty acids.

*At the time—two decades before Calvin started to work on the metabolic pathways of photosynthesis—circular metabolic pathways were a revolutionary concept. Thus, Krebs's original publication on the citric acid cycle, which was to earn him the 1953 Nobel Prize in Medicine and Physiology, was rejected outright when he submitted it to the journal *Nature*.

The cycle produces chemical energy in the shape of the molecules ATP, NADH, and $FADH_2$. Adenosine triphosphate is a ribonucleotide, which we first met in chapter 6 as the energy "currency" of the cell and a potential vestige of the RNA world. Reactions that produce metabolically useful energy almost always produce it in the form of ATP, and metabolic processes that consume energy almost always use the energy stored in ATP. NADH and $FADH_2$ (both also ribonucleotides—more vestiges of the RNA world?) are the "reduction currency" of the cell, delivering reducing power to any reaction that needs it. One such reaction is the reduction of oxygen to form water in a process (respiration) used for the production of still more ATP in the electron transport chain.

Even a quick glance at the numbers reveals the advantage that the Krebs cycle provides to species that adopt an aerobic lifestyle. The anaerobic metabolism of glucose using glycolysis alone—that is, splitting glucose into two molecules of pyruvate—produces only 2 ATP. In contrast, the Krebs cycle followed up by the electron transport chain squeezes 24 ATP out of each and every molecule of glucose. In combination with glycolysis and the decarboxylation of pyruvate to acetyl-CoA, aerobic metabolism produces 36 ATP per glucose, achieving an eighteenfold increase in energy yield over the anaerobic glycolysis alone.

Because of its tremendous efficiency, it is not surprising that the aerobic lifestyle evolved and spread soon after the atmosphere became oxic. Like photosynthesis, the Krebs cycle was invented by bacteria, some of which later joined forces with other cells in an odd form of intracellular symbiosis. With the advent of this fusion, a much higher level of metabolism became possible, and with that came the possibility of *multicellular organisms.* But first, evolution had to invent a more complex class of single-celled organisms.

Eukaryotes: Bigger and Better Cells

Evolution cannot display foresight or goal-oriented strategic planning. But it is clear from the history of life on our planet that the invention of oxygen-producing photosynthesis paved the way for evolution of more complex life, a niche that was taken over in its entirety by the first eukaryotes and their descendants.

As we humans, along with all the animals and plants that end up on our dinner plates (and the yeasts that produce the alcoholic drinks in our glasses), are eukaryotes, we tend to take the cell nucleus for granted. However, it was certainly not written in the stars that life must be eukaryotic. After all, when life first evolved on Earth, and for as much as 1.7 billion years afterward (over a third of our planet's history), prokaryotic life forms were the only game in town. And while these may have produced some fairly complex higher-order structures, such as the stromatolites we discussed above, evolution had to invent multicellularity before life could diversify into spectacular forms and achieve whole new levels of complexity.

Look at cells through a microscope and you will be able to tell whether they are eukaryotes or not. Typically, our cells and those of most other eukaryotes are ten times larger in each dimension than the simpler prokaryotes. Peering down a light microscope, this makes the difference between seeing a cell with internal structure and just seeing a dot.

The defining difference that gives eukaryotes their name (again, *eukaryote* means "true nucleus") is that their genetic material is isolated from the rest of the cell in the nucleus, surrounded by a double membrane that resembles the membranes of bacteria (fig. 7.3). The DNA in eukaryotes is typically organized in several, long linear units, the chromosomes, whose coiling and packaging is usually controlled by histones, a class of proteins that does not exist in bacteria (but does occur in some archaea). The synthesis of messenger RNA (transcription) also takes place in the nucleus, while protein synthesis is carried out by ribosomes in the cytoplasm. This separation creates additional logistics problems and seems rather troublesome at first glance. So what is the evolutionary advantage that enticed eukaryotes to keep the DNA wrapped up? The clue may lie in the additional editing of messenger RNA that is made possible by this separation. In bacteria, the front end of a messenger RNA can go into the translation machinery while the rear end is still being transcribed. In eukaryotes, in contrast, the messenger RNA is fully synthesized in the nucleus, where it can then be "edited" in various ways before finally being transported to the cytoplasm, where it is translated to make proteins. While making matters more complicated, this separation in space and time allows the cell to introduce additional mechanisms to control gene expression, which in turn sets the stage for vastly more complex organisms.

Another important way in which the eukaryotic cell stands out is the presence of many other membrane-bound compartments such as mitochondria, chloroplasts, lysosomes, and the endoplasmic reticulum. Each of these compartments brings in a complete set of important functions. Several of them, most notably the mitochondria and the chloroplasts (in charge of aerobic metabolism and photosynthesis, respectively), are thought to have arisen from formerly independent bacterial symbionts, which, over the generations, became better and better integrated into the host cell and lost their independence along with most of their genes. And

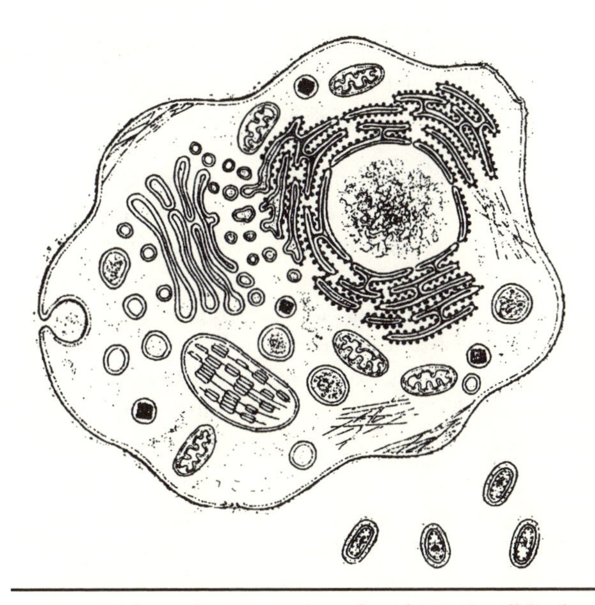

FIG. 7.3. Schematic cross section of a eukaryotic cell (a plant cell is shown here) in comparison with a bacterium. Note that eukaryotic cells tend to be an order of magnitude larger in each dimension, so their volume exceeds that of a bacterium by around three orders of magnitude.

this, when you think about it, brings us to the question: where did the eukaryotes come from, anyway?

Comparison of protein and gene sequences has enabled researchers to trace back the family tree of life much more precisely than would have been possible based on outward appearance (phenotype) alone. At first, they studied enzyme sequences, but found their efforts limited by the effects of convergent evolution (remember: similar environmental requirements can lead to similar adaptations at both the organismal and protein levels), which can produce similarities that do not imply relatedness. As described at the beginning of this chapter, Carl Woese focused his sights on ribosomal RNAs and came to the conclusion that, whereas living things had previously been divided only by the presence (eukaryotes) or absence (prokaryotes) of a nucleus, the oldest and deepest division between species separates the tree of life into three main branches. In the decades since, researchers have switched to the analysis of genes, which, unlike even rRNA, contain variability that is under almost no selective pressure at all—such as a change in the third base of codons where the third base is redundant, which doesn't change the amino acid sequence of the encoded protein (chapter 6). The ever-increasing numbers of protein, gene, and genome sequences that have become available have provided more and more convincing evidence in favor of the tripartite tree of life.

The three domains of life are fundamentally different in many ways. In some aspects, there are resemblances between two of them that exclude the third (e.g., only Bacteria and Eukarya synthesize fatty acids, only Archaea and Eukarya wrap their DNA around histones, and both Bacteria and Archaea lack nuclei), but there is no case for a grouping into two domains any more. The very last doubts about that were removed by the first complete genome sequence of an archaeon (*Methanococcus jannaschii*), which was published in 1996 and illustrated conclusively that the Archaea are no more closely related to the *Escherichia coli* living in our guts than they are to us.

However, even the most modern analytical methods have failed to answer two important questions: first, how do the three largest branches on the tree of life relate to each other; and second, where is the root? Comparisons based on different genes or groups of genes yield very different answers to these questions. From these contradictions it seems increasingly likely that "vertical" descent of species from earlier species, along the direct lines of a family tree, does not account for the whole story of life on Earth. Exchange of genes between separate species living at the same time must have played an important role.* This kind of horizontal gene transfer can still be observed among microbes, for example when researchers study the spread of genes conferring antibiotic resistance or other crucial survival skills. It became obvious that, given the abundant occurrences of horizontal gene exchange in the history of life, the attempt at drawing simple family trees relating all living species to a smaller set of ancestors and ultimately to a common root was destined to fail.

Recently, Maria Rivera and James Lake at the University of California, Los Angeles, delivered a different description for the crucial early phase of evolution when they applied a new set of algorithms to its modeling. They compared the genomes of ten organisms representing all three domains of life, using an algorithm ("conditioned reconstruction") designed to cope with both horizontal and vertical gene transfer without discrimination.

*Gene exchange also weakens our arguments that any traits that are shared across all three domains of life must have been inherited from LUCA; some of the shared traits could have arisen in one branch, after LUCA, and been horizontally transferred to the others. Still, most authorities seem to agree in broad detail with the description of LUCA we've given here.

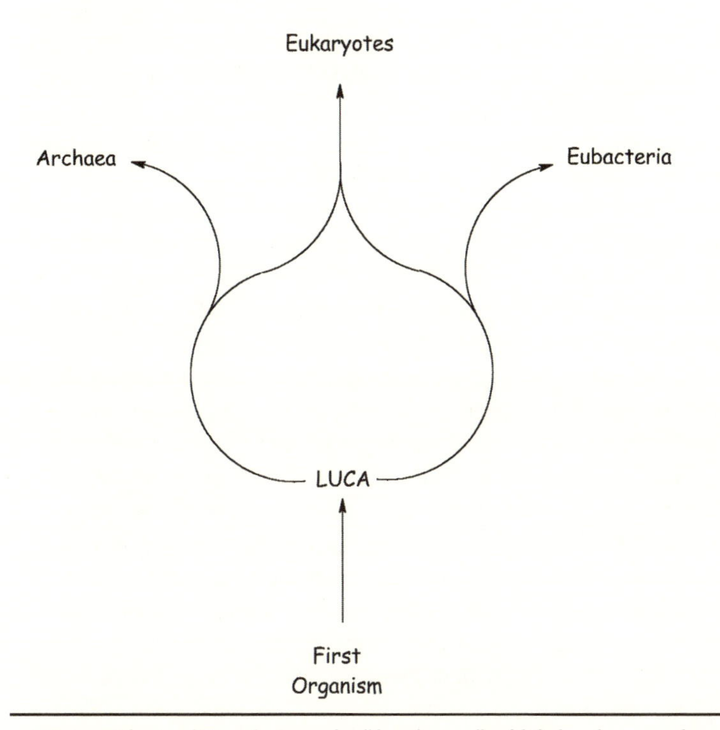

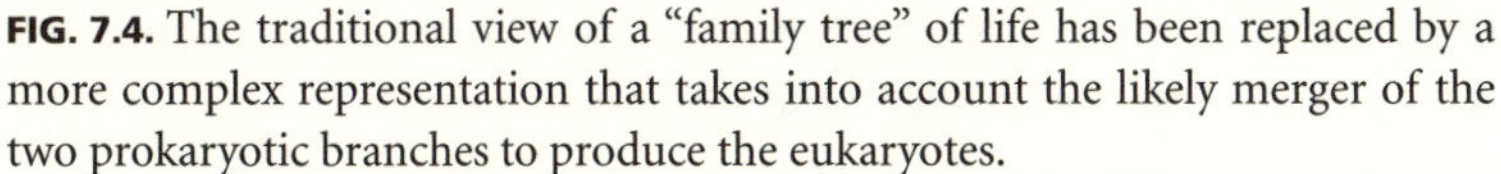

FIG. 7.4. The traditional view of a "family tree" of life has been replaced by a more complex representation that takes into account the likely merger of the two prokaryotic branches to produce the eukaryotes.

In this method, one of the genomes (the conditioning genome) is picked as a standard that does not enter the resulting tree, as it serves as a reference point for the others. Thus, there is a simple built-in control: one can construct trees based on different choices of conditioning genome and then overlay them. Rivera and Lake first tested this method on the parts of the prokaryotic family tree that are already well-described, then applied it to the question of how the deepest branches—the three domains of life—relate to each other. Sifting through the results with the highest statistical significance parameters, they realized that all of them were permutations of a single pattern that can best be described as a ring (as shown in fig. 7.4).

Biologically, this finding implies that Eukarya arose from both Bacteria and Archaea, possibly via the fusion of two early cells. This interpretation is consistent with earlier results of genome comparisons showing that eukaryotic genes in charge of information processing (ribosomes, translation factors, enzymes of DNA processing) are more closely related to their archaeal counterparts than to the bacterial ones, while the reverse is true for the genes related to metabolism. With several independent studies now confirming a "mixed" origin of eukaryotes, it seems almost certain that they arose from some kind of marriage between the two older, more "primitive" domains.

When did all this happen? That's not so easy to say, in part because, from the outside, the earliest eukaryotes probably didn't look that different from contemporary prokaryotes. Worse, single-celled organisms don't leave much of an impression in the fossil record. Because of these difficulties, the dates cited for the oldest physical (as opposed to molecular) fossils of eukaryotes vary between 1.7 and 2.1 billion years ago. Remarkably, though, late Archean shales have been found in northwestern Australia that contain steranes, large organic molecules that are thought to arise from the degradation of sterols such as cholesterol. And while a few bacteria incorporate sterols into their membranes, no known prokaryotes are capable of synthesizing the complex 28-carbon sterols that must have given rise to the steranes found in these rocks. Indeed, at least on the contemporary Earth, *only eukaryotes are known to produce these molecules.* Thus it seems that a key attribute of eukaryotic biochemistry, if not eukaryotes themselves, had evolved by 2.7 billion years ago—even though, at that stage, it was far from obvious that they would one day rise so far above the lifestyle of their primitive ancestors and start to write and read books about astrobiology.

Stepping up to Multicellular Life: Explosions and Extinctions

The invention that really set eukaryotes apart from the crowd was the step to multicellular organisms. Green algae were among the pioneers of higher organization in colonies, but didn't immediately make the transition to developing a body plan. Animals took the lead here, gradually evolving from very primitive forms with just two layers of cells to three layers, three layers with a cavity, and onward to worms. The first clear record of this evolution does not arise in the rock record until only 600 million years ago. This is remarkably recent in geological terms. After our planet became habitable some 3.8 billion years ago, it apparently took only a few hundred million years for single-celled life to arise. Multicellular life, in contrast, took at least an order of magnitude longer, suggesting that it was not an obvious transition that any old bacterial species could have made. It seems several lineages of single-celled organisms had to come together to form the more sophisticated eukaryotic cells. This might also have required the advent of a dense, oxygen atmosphere (see sidebar 7.2).

Why did bacteria fail to move upward? Cyanobacteria, for example, are amazingly sophisticated, having mastered photosynthesis, nitrogen fixation, symbiosis with fungi to form lichens, and even developing circadian clocks. So why did some obscure eukaryote steal the limelight from them, increasing the complexity of living beings and inventing the higher plants and animals? There are clearly many contributing factors, no doubt some of which we still don't understand. But a few of the features that enabled the jump are clear: the organization of space, organization of the genome, and sex.

The most striking difference that distinguishes eukaryotic cells from bacteria is not just the volume of their interior space (typically 1,000-fold larger), but the way this space is organized in eukaryotes into compartments of well-defined function. Bringing together the information processing of the nucleus, the aerobic metabolism of mitochondria, possibly the photosynthesis of chloroplasts, and other functions in separate entities, even a formally single-celled eukaryote is effectively a "multicellular" organism.

More importantly, eukaryotes abandoned the single circular DNA that bacteria use and invented linear chromosomes. Even the humble, single-celled eukaryote baker's yeast has sixteen separate linear chromosomes. While there is a limit to how much DNA you can store and process in a ring without ending up in a lethal tangle, the organization into chromosomes offered not only more storage space but a natural way of expanding the space, namely by adding new chromosomes. Thus, the number of chromosomes varies widely between different eukaryotic species.

Most importantly, the new style of genome organization enabled eukaryotes to embark on a completely new way of fostering genetic diversity, while ensuring the genetic stability that keeps a species together. This wonderful new tool of evolution, invented around a billion years ago, is known as meiosis on the cellular level, but on the organism level it is called sex. Biologists have argued about its usefulness. In comparison with an asexual reproduction mechanism, where every individual can have offspring, the maintenance of a nonreproductive gender (e.g., the human male) is a complete waste of energy, one might argue. However, the success of sexual reproduction throughout the animal kingdom and in much of the world of plants shows that the benefits to the species more than compensate for this loss.

So, equipped with these advantages, some protists finally got their act together and became the founders of zoology, sometime before 600 million years ago. We know very little about their first efforts, as they don't show up in the fossil record very clearly, and molecular analyses have not yet resulted in a convincing reconstruction of the first animal.

SIDEBAR 7.2

Weighing the Probabilities

The evidence—albeit weak—that life may have taken only tens of millions of years to both arise and proliferate after the end of the late heavy bombardment is often taken to imply that the formation of life is a very likely event. But this logic is as flawed as any statistics based on only one sample. This logic is further undermined if rapidity is an imperative for the formation of life, and speed may well be an imperative. Miller-Urey chemistry absolutely requires a reducing atmosphere and, due to the loss of hydrogen via photolysis, the Earth's atmosphere started to oxidize from the day it was formed. This oxidation was sufficiently rapid that the earliest rocks that record evidence of the Earth's surface suggest it was fairly oxidized by the time they formed. Thus it is possible that life on Earth captured an extremely narrow window of opportunity between the end of sterilizing impacts and the oxidation of the primordial atmosphere.

In contrast to the potentially rapid origins of life on Earth, though, the development of complex cells, the step from complex cells to complex organisms, and the step from complex organisms to intelligent life proceeded at a much more leisurely pace. But does the fact that it took 4 billion plus years for intelligence to arise on Earth imply that it requires so long everywhere? Or are we just slow? The answer to this question is at least a qualified "it takes time." For example, intelligence no doubt requires multicellularity (or, probably harder, the equivalent complexity in a unicellular organism). This level of complexity, in turn, probably requires

But these events laid the groundwork for something big, because only 50 million years later the fossils document a sheer explosion of animal diversity.

Now isn't that ironic? You wait some 3 billion years for animals to come along, and then they all arrive at once. Or so it may seem. Some 650 million years ago, several billion years after the first traces of life on Earth, there were still no traces of animals in the fossil record, and yet, 200 million years later, they were everywhere. In fact, within 200 million years of the first animal in the fossil record, there were more than a hundred different orders of animals, almost as many as today.*

Essentially, during the Cambrian Period (542–488 million years ago) and the subsequent Ordovician Period (488–443 million years ago), animals tried out many different body plans, many of which are still in use.

*An *order* is a phylogenetic group that is larger than a *genus*, smaller than a *class*. For example, mammals are a class (Mammalia), in which Primates (e.g., ourselves), Rodentia (e.g., mice), and Proboscidea (the elephants, of course) are orders.

the very active metabolism provided by oxygen. And it probably does require billions of years for an oxic environment to form. The reason is that, for the first billion or so years, all of the oxygen produced by photosynthesis (and abiological photolysis) is consumed by the oxidation of rocks. Only after enough reducing material has been removed from the system (carbon in the form of oil, coal, and so forth, or the hydrogen lost to space) can a net flux of oxygen be achieved. On Earth this did not occur until our planet was middle aged. When it finally did occur, it took less than 200 million years for multicellular organisms to evolve into every current phylum (and many that are long extinct) and another 600 million years for us latecomers to arrive on the scene.

So what, then, were the chances that higher life would evolve on the Earth, culminating (or so we like to think) in intelligent life? This is, of course, impossible to determine. But that doesn't mean we can't have some fun speculating.

A potential important input to this speculation is that intelligence didn't arrive here until the planet was 4.56 billion years old, and took at least 2.7 and possibly as long as 3.8 billion years to show up after the origins of life itself. These time periods, which are long even when compared with the 13.7-billion-year age of the Universe, suggest we are a lucky break. If it really requires that much time to evolve from the first, simplest organisms to something smart enough to read a book about astrobiology, what are the chances that something won't come along and kill life off before it gets there? Given the probability of sterilizing impacts and the certainty of moving habitable zones (chapters 2 and 3), it is not at all a trivial thing for a planet to remain habitable for a quarter of the age of the Universe.

This unrivalled burst of evolutionary inventiveness, known as the Cambrian explosion, has mystified biologists from Darwin to the present. Today there are two fundamentally different schools of thought on this issue, each with its own toolkit of possible explanations and interpretations.

The "late arrival" school, represented by Stephen Jay Gould (1941–2002) of Harvard University, maintains that what we see in the fossil record is essentially what happened, and the diversification did indeed take place unusually fast, creating as many as fifty new orders in just 10 million years. One possible explanation sees the explosion as an arms race triggered by the use of biomineralization by animals. The controlled deposition of minerals from body tissues enabled animals not only to develop skeletons (which also allowed them to diversify into more complex shapes and larger sizes), but also to grow claws, rasps, and teeth with which they could prey on other animals. Predatory lifestyles opened up additional ecological niches, and triggered defensive measures, such as biomineralized shells, in the animals threatened by them.

Similarly, the development of eyes may have further intensified this interspecies arms race.

Of course, paleontologists can study only animals that leave fossils, and thus the impression of an "explosion" is made even more dramatic by the fact that mineralized tissues such as bones, teeth, and shells have a much better chance of being preserved than the soft tissues of the animals that went before the onset of biomineralization. This argument leads us to the second school of thought, the "early arrival" model (favored by Darwin), which claims that animal diversity existed for hundreds of millions of years before the Cambrian Period, but didn't show up in the fossil record because either the animals were "too soft" to fossilize properly or the conditions were unfavorable for their preservation. Extreme versions of this view place the origins of animal diversification as far back as 1.2 billion years ago.

Improved recognition of the fossilized traces of soft animals and "molecular clock" studies that date divergences by the slow ticking of mutations suggest that some animal diversity did indeed exist before the explosion. Nevertheless, most researchers would not allow more than 150 million years for this hidden period of animal evolution. In a compromise between early- and late-arrival theories, some describe this period as the "fuse" that eventually triggered the Cambrian explosion. Several authorities place one of the deepest divisions in the family tree of multicellular life, the one between protostomes (including mollusks, insects, crustaceans) and deuterostomes (including us vertebrates), at 670 million years ago, which would allow the "fuse" some 125 million years to burn.

Some of the important changes happening during this "fuse" period might have been invisible not just to the fossil record but even to a time-traveling biologist inspecting the actual animals. Genes that act on a higher level of organization, namely by regulating the expression of other genes during embryonic development, such as the *Hox* genes, were an important prerequisite for the evolution of complex body plans. *Hox* stands for "homeotic complex," a set of regulatory genes originally discovered in the fruit fly *Drosophila,* and then also in vertebrates. It governs the segmentation of animals during embryonic development. The homeotic complex of humans as well as mice contains thirty-nine genes organized in four clusters. The genes tend to be redundant, such that knocking out one of them often has only minor effects, while combination knockouts can seriously disrupt development. This redundancy probably played an important role in the diversification of body plans during the Cambrian explosion. Possibly these genes developed and diversified in groups of animals that looked unassuming for millions of years, but quietly built up the genetic machinery that made the explosion possible.

During the Cambrian explosion, animals developed new lifestyles and thus populated new ecological niches. In the following periods, by contrast, the total number of animal orders did not change significantly. When new periods of inventiveness did occur, they were typically preceded by mass extinctions. For example, when the dinosaurs disappeared, they left vacant ecological niches that were subsequently filled by the newly ascendant mammals. Geographical change such as the breakup of the supercontinent Gondwana, some 180 million years ago, promoted biodiversity at the level of species, genus, and family but not, however, on any larger scale.

Thus it is obvious that many different factors, from genetic to ecological, and from new inventions to mass extinctions, were crucial for life to attain its current global coverage and high biodiversity, not to mention the production of a human civilization. We shall have to take these factors into consideration when we discuss the probability of extraterrestrial life and civilizations in chapter 10.

Conclusions

It's been a few chapters since we've said it, and so it probably bears repeating: biology is a provincial science. Given that all life on Earth arose from a biochemically complex common ancestor, it's not so easy to figure out which aspects of our biochemistry and cell biology reflect adaptations to the fundamental issues related to life on a terrestrial planet, and which are merely historical artifacts of evolutionary chance. But then again, we biologists have to play the cards we are dealt. From that perspective, detailed studies of the evolution of life on Earth are probably the best approach we have to understanding how life is defined, constrained, and encouraged by the physical reality of growing up on a small, rocky planet.

So what does this ultracompressed history of life teach us in the context of astrobiology? The evidence that life existed by 2.7 billion years ago is rather strong, but this is nearly half the age of the Earth and a full 1.1 billion years after the end of the late heavy bombardment. The evidence for life in older rocks is progressively weaker, and very weak indeed for the Earth's oldest rocks. If this tenuous evidence holds up, and life was common at 3.8 billion years ago, this means it arose reasonably fast—within 100 million years of the end of the late heavy bombardment (see sidebar 7.2). More complex eukaryotic cells, however, did not arise until at least a billion years after the formation of life itself, and the birth of multicellular life forms required the formation of an oxic atmosphere, which took another billion plus years after that. On Earth, at least, a reasonable fraction of the age of the Universe went by before an obscure, mammalian branch of the eukaryotic tree of life evolved to a point where you could be reading this paragraph.

Further Reading

Earliest life/evolution of life/origins of eukaryotes. Knoll, Andrew. *Life on a Young Planet.* Princeton, NJ: Princeton University Press, 2003; Fortey, Richard. *Life.* New York: Vintage Press, 1997.

Evidence for life at 3.8 billion years. Fedo, C. M., and Whitehouse, M. J. "Metasomatic origin of quartz-pyroxene rock, Akilia, Greenland, and implications for Earth's earliest life." *Science* 296 (2000): 1448–52; Sano, Y., Terada, K., Takahashi. Y., and Nutman, A. P. "Origin of life from apatite dating?" *Nature* 400 (1999): 127; Dalton, R. "Fresh study questions oldest traces of life in Akilia rock." *Nature* 429 (2004): 688.

Evidence for life at 3.5 billion years. Schopf, J. William. "Microfossils of the Early Archaean Apex chert: new evidence of the antiquity of life." *Science* 260 (1993): 640-46; Brasier, M. D., Green, O. R., Jephcoat, A. P., Kleppe, A. K., Van Kranendonk, M. J., Lindsay, J. F., Steele, A., and Grassineau, N. V. "Questioning the evidence for Earth's oldest fossils." *Nature* 416 (2002): 76–81.

Origins of photosynthesis. Gross, Michael. *Light and Life.* Oxford: Oxford University Press, 2003.

CHAPTER 8

Life on the Edge

In the spring of 1977, the geologists John Corliss of Oregon State University and John Edmond (1944–2001) of the Massachusetts Institute of Technology boarded the research submarine *Alvin* for humans' first firsthand look at a mid-ocean ridge. They were following up on observations made two years earlier in the Atlantic that these ridges—a globe-girdling chain of mountains beneath the sea—seemed to consist of freshly solidified basalt, suggesting that an active source of lava was nearby. The researchers were on a hunt for "spreading centers" where new crust is formed. The existence of such centers was predicted by plate tectonics, a theory that was first proposed by the German geophysicist and meteorologist Alfred Wegener (1880–1930) back in 1912 but was only recently beginning to achieve widespread acceptance.*

When *Alvin* reached the slope of the ridge, some 2,500 meters beneath the surface of the Pacific, the geologists noticed that the outside temperature was five degrees higher than the normal 2°C of the ocean's depths. At the time, marine geologists had theorized that the entire volume of the oceans somehow flows through hot volcanic rocks every 8 million years or so—only this could account for the chemical composition of seawater, which is drastically different from that of river water boiled down in an evaporation pan—but no one had yet identified the hydrothermal features that might account for the cycling. Hence, this hint of hot springs on the ocean floor was already a sensational discovery for the researchers in *Alvin,* and they excitedly took samples so that they could later determine the chemical composition of this unexpectedly warm water. Still excited, they piloted *Alvin* up to the top of the ridge, where a much bigger sensation was waiting for them. Where they had expected to find a stark "desert" of bare, lifeless basalt, freshly erupted from the spreading center atop the ridge, they found an oasis 100 meters in diameter, with warm water sifting through every little

*More precisely, Wegener proposed continental drift, based on the close fit of eastern South America with the western coast of Africa and coincident mineral formations on either side of the Atlantic, but his theory did not provide details as to why or how the continents might be moving. Plate tectonics, per se, was developed in the 1950s.

SIDEBAR 8.1

Stress Responses

Adaptation on evolutionary timescales is one way of responding to extreme environmental conditions. But given that not all habitats are equally stable, some organisms can temporarily find themselves in "hot water" on much shorter timescales. Thus the ability of organisms (both mesophiles, those of us who live under "normal" conditions, and extremophiles) to survive short-term deviations from the conditions they've evolved to live under is also a fundamental factor in defining the range of conditions under which life can survive.

Here on Earth, the most important and universal form of this response to stress is the expression of specific stress proteins, of which the heat shock proteins (Hsps) are among the best studied. This is a large family of proteins that are expressed (produced) in cells in response to any unaccustomed rise in temperature. The heat shock response had been known and characterized as a phenomenon of gene regulation decades before researchers understood the primary functions of these proteins. Then, in the late 1980s, it was established that several of the main components of the heat shock response act as molecular chaperones—that is, they protect other proteins, which are newly synthesized or partially unfolded, from intermolecular interactions that would favor aggregation (which leads to loss of function) over correct folding.

The "classic" chaperones DnaK, DnaJ, GroEL, and GroES, which in *Escherichia coli* are combined in a complex, efficient protein-processing pathway, were studied intensely in the 1990s. Over time, a range of additional functionalities were discovered in other heat shock proteins. For example, the major component of vertebrate eye lens, alpha crystallin, has a chaperone function and is related to the small Hsps (it is presumably there to prevent the aggregation of lens proteins, which if left unchecked would lead to cataracts).

crack of the seafloor, and richly populated with clams, crabs, sea anemones, and large pink fish. As Edmond later recalled in *Scientific American,* they spent the five remaining hours of their dive in frantic excitement. They measured temperatures, conductivity, pH, and oxygen content of the seawater, took photographs, and collected specimens of all the animal species.

Holger Jannasch (1927–98), a German marine biologist working at the Woods Hole Oceanographic Institution, was one of the first to hear the news. He later recalled that he "got a call . . . from the chief scientist, who said he had discovered big clams and tube worms, and I simply didn't believe it. He was a geologist, after all" (quoted in *Time Magazine,* August 14, 1995).

Several other Hsps take part in the destruction of discarded proteins, and at least two (Hsp31 and DegP) can switch between the functions of chaperone and destroyer. In cooperation with other, more conventional chaperones, Hsp104 can even rescue proteins from aggregates (the biochemical equivalent of "unboiling an egg"). Hsp90 specifically chaperones transcription factors (proteins that regulate gene expression), folding them even if they are mutated and thus helping to silence the effects of mutations. But when an acute heat shock requires Hsp90 for use elsewhere as an emergency response, it stops chaperoning the transcription factors. This allows the "expression," as it were, of previously silent mutations in these factors, which could result in a spectacular increase in developmental variation. This finding suggests that Hsp90 has an important role linking environmental stress to the generation of new biological functions.

Responses to other kinds of stress typically involve a combination of heat shock proteins and proteins more specific to the given type of stress. Among the better-studied examples of non-heat-related stress proteins are the cold shock proteins. The prototype, CspB from *Bacillus subtilis,* is known to serve as an RNA chaperone in that it keeps mRNA from folding into loops that might inhibit its translation. Only limited information is available on proteins specific to other kinds of stress.

Stress proteins are universally present in all organisms we know and are thus presumed to be ancient in evolutionary terms (the ability to survive temporary environmental changes provides a significant selective advantage). Moreover, their involvement with key regulatory proteins suggests they have played an important role when stressful environmental changes made relatively rapid adaptation necessary. These stress proteins might therefore hold additional clues to the questions of how life on Earth managed to adapt to some surprisingly unstable habitats.

The Art of Living Dangerously

Living organisms tend to be sensitive to drastic changes in their environments. Heat and cold, pressure, drought, salinity, acids and bases—all disrupt the crucial interactions that keep biomolecules folded and functional, and quickly put an end to the fragile state of chemical disequilibrium we call life. Therefore, scientists have tended to assume there are strict boundaries to the biosphere, imposed by Terrestrial life's requirement for a rather narrow and specific range of physical conditions (see sidebar 8.1).

Discoveries in the past few decades, however, have shown that life isn't always as sensitive as we might have imagined, and that the limits of life on Earth are far from well defined. Historical notions of what is, or is not, a hostile environment have turned out to be erroneous. Methods routinely used for sterilization, including boiling, freezing, and γ-ray treatments, turn out to be deadly for most—but not all—microbes. For every extreme physical condition investigated, extremophilic organisms have shown up that not only tolerate these conditions but often even require them for their survival. With these discoveries, the expanse of the known biosphere has grown and the putative boundaries of life have expanded. The last thirty years, in particular, have witnessed substantial shifts in what scientists consider the limits of habitable environmental conditions.

While there are certain hard physical limits to the existence of DNA-based cellular life, these limits are far from the normality of common or garden organisms such as *Escherichia coli* and *Homo sapiens* (a normality admittedly defined by anthropocentric thinking!). Here we take a brief look at some of the extreme conditions faced by organisms living on our planet, considered under the fundamental astrobiological question of what these findings tell us about the prospect of finding life elsewhere in the Solar System and in the wider Universe.

Thermophiles

Microbial activity at temperatures above the "normal" range of 20°–40°C had been reported in the nineteenth century, but the upper limit of life's known temperature scale has risen rapidly over the past four decades (fig. 8.1). An important foundation was laid when Thomas Brock started cultivating heat-resistant bacteria in the hot springs of Yellowstone National Park in the 1960s. His discoveries included *Thermus aquaticus*, which two decades later was to become the first and most common source of heat-stable DNA polymerase for the polymerase chain reaction—now an essential, everyday tool in the biotech industry and all across the life sciences (see sidebar 8.2). The temperature records that Brock's organisms set, however, were not destined to last, as even more hostile habitats remained to be discovered.

Two years after Corliss and Edmond and their team discovered warm springs on the seafloor that hosted a surprisingly rich ecosystem, another *Alvin* expedition found the first hydrothermal vents. These extremely hot (up to 350°C) springs erupt with great force from chimney-like deposits rising several meters above the seafloor. (Under the elevated hydrostatic pressure at such depths, the boiling point of water is raised considerably; at 2000 m, for example, it is 340°C.) The drastically increased solubility of certain minerals in the hydrothermal fluid leads to instant precipitation when the mineral-rich fluid mixes with cold seawater, producing both the characteristic chimney walls and the "smoke plume" that earned these vents the nickname "black smokers" (fig. 8.2).

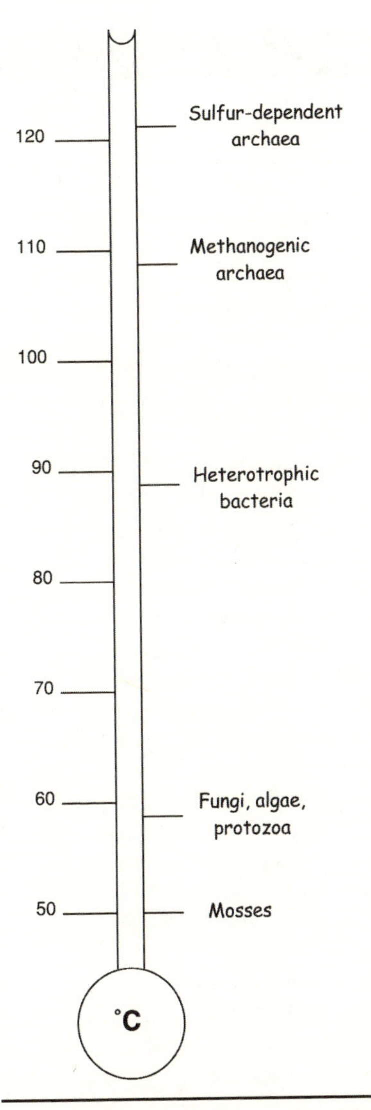

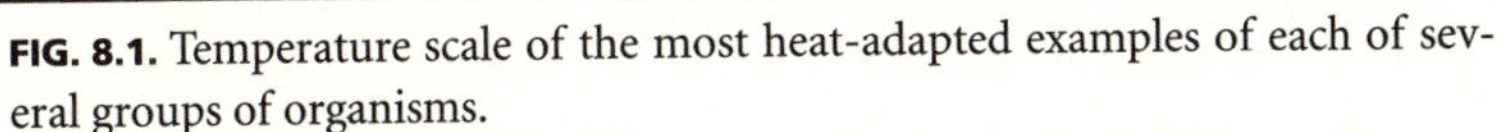

FIG. 8.1. Temperature scale of the most heat-adapted examples of each of several groups of organisms.

Around these springs, complete ecosystems with complex food webs flourish without daylight or any carbon source more exotic than carbon dioxide.

Detailed investigation of the ecology of hydrothermal vent communities revealed that at the base of their food chain are single-celled chemotrophs, organisms that live by taking in abiological nutrients and

SIDEBAR 8.2

Commercial Interest in Extremophiles

Because conditions that count as extreme for biologists are often fairly standard in industrial settings, astrobiologists aren't the only people interested in extremophiles. Many processes in the chemical industry, for example, are routinely run at high temperatures and pressures. The food industry uses all kinds of extreme conditions to exclude food pathogens. So it is not surprising that the discovery of extremophiles spawned a certain degree of interest from industry.

Enzymes from thermophilic organisms, which can be used in processes that combine traditional high-temperature protocols with enzymatic reactions, have proved of particular interest to the chemical and pharmaceutical industries. Agro-tech has used genetic engineering to transfer frost resistance from one organism to another. And the increasing understanding of the survival strategies used by stressed microbes can help to optimize sterilization processes. For instance, *Bacillus* spores are known to survive high-temperature treatment (e.g., in a pressure cooker), but certain combinations of temperature and pressure changes can trigger the spores to germinate, and as they give up their enhanced protection they can be destroyed more efficiently.

The single most successful commercial product derived from an extremophile is the thermostable DNA polymerase used in the polymerase chain reaction (PCR). This process revolutionized molecular biology in the 1980s, and it was the discovery of DNA polymerases that remain stable at 90°C, the temperature needed to fully dissociate a DNA double helix, that made it possible. The original protocol developed by Kary Mullis (which won him the Nobel Prize in Chemistry in 1993) involved the polymerase of *Thermus aquaticus,* now known as Taq polymerase. In the 1990s, however, the rapidly growing competition in the PCR field made available a range of enzymes derived from thermophiles, including *Pyrococcus furiosus* (Pfu).

This goes to show that there can be immense benefits from research into the apparently "offbeat" areas of science such as survival under extreme conditions. Further discoveries in this area will certainly prove useful for research, medical applications, and indeed everyday life. One day, our knowledge of extremophiles' survival strategies might even help us to brave the extreme conditions on other planets ourselves.

using them as the raw materials from which they build their cells. These single-celled organisms, in turn, live in a close, mutual symbiosis with, or are eaten by, higher eukaryotes such as clams and the now well-known, multimeter-long tube worms. These macroscopic, multicellular organisms, however, live in the much cooler water centimeters to meters

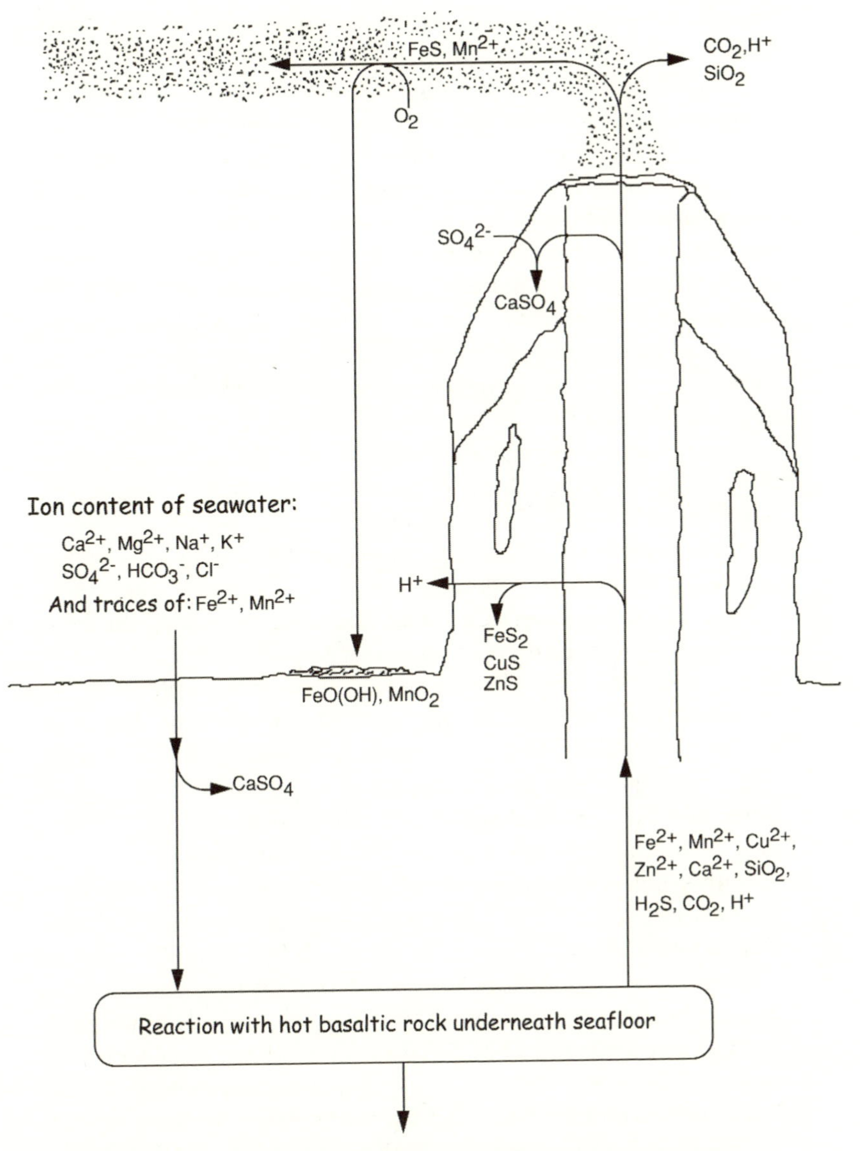

FIG. 8.2. Schematic cross section of a hydrothermal vent, or black smoker, showing the reaction paths of the most important minerals. Even before black smokers were discovered, geologists had predicted that such reactions must be occurring at the ocean floor, in order to account for the unusual salt content of seawater. It is estimated that the entire water volume of our oceans runs through black smokers once every 8 million years.

away from the vents and thus are not themselves particularly thermophilic.

The ability to thrive at high temperatures is not the only remarkable thing about the vent communities. Unlike the biosphere we humans know and love, the vent communities are not dependent on photosynthesis for their reduced carbon. Instead they use the reduction potential available in sulfides extracted from the crust by the hot vent water to reduce carbon dioxide to the amino acids, nucleic acids, and other organic materials they need for growth. Contrary to many claims, however, the vent organisms are just as dependent on the Sun as we humans are; the reason sulfides are able to provide the energy necessary for life is that a mixture of sulfides and oxygen is out of equilibrium relative to sulfate and water. Thus the chemical disequilibrium on which the entire vent ecology is founded relies on the availability of molecular oxygen, which is produced by photosynthesis. Without the downward diffusion of photosynthetically produced oxygen to the vents, they would likely be too close to chemical equilibrium to provide enough energy for anything but the simplest of living communities.

The adaptation to survival at high temperatures is closely linked to the chemical needs of these organisms, specifically to the source of reducing power they use. They require large amounts of sulfides, which are provided by the seafloor springs and hydrothermal vents, but which tend to be insoluble at the much colder temperature of bulk seawater (2°C). The organisms are therefore able to harvest more sulfide the closer they live to the hot spring, and this produces a strong selective pressure for the evolution of extreme thermophilic properties.

Other hot environments that have also yielded extremely thermophilic microbes include solfatare fields, which are volcanic soils permeated by hot vapors found in such places as Yellowstone National Park and sites in Iceland and Sicily. Surface waters in solfatare fields are often near the boiling point and rather caustic as well, with pH values ranging from an extremely acidic 0.5 to a moderately basic 9 (more on environmental pH later in the chapter). They also usually contain a diverse mix of sulfur compounds. Deeper layers of solfatare fields are typically less acidic, and are anaerobic. Under these reducing conditions, the sulfides of heavy metals tend to precipitate, coloring the soils a gooey black. Some of the most extreme thermophilic and acidophilic microbes have been isolated from these environments.

A number of thermophilic organisms have been studied in detail, several of which are of particular interest to us astrobiologists. These include two of the most thermophilic bacteria, *Aquifex pyrophilus* and *Thermotoga maritime*, which, by the way, seem to be among the most "ancient" of all prokaryotes. That is, they seem to be the living organisms most closely related to the putative branch organism from which Archaea, Eukarya, and Bacteria diverged.* The 80°C optimal growth temperature and the 89°C and 96°C (respectively) maximum growth temperatures of these two types of bugs, however, are surpassed by other, more recently characterized species. These include *Pyrolobus fumarii,* an archaeon isolated from a black smoker with a growth range of 90°–113°C, and the current record holder, an as yet unnamed strain designated "strain 121." This species, which grows at 121°C and can survive several hours at 130°C, cannot grow at all below 85°C. It is an anaerobe and seems to make its living by extracting electrons from simple reduced compounds in thermal vents and using them to reduce Fe^{3+}.

Life under conditions of extreme heat, especially when it is coupled to low pH, requires that an organism's proteins remain folded and active in these seemingly challenging environments. Most eukaryotic proteins, for example, unfold at modest temperatures or modest pH, as is seen when you boil an egg or when a Mexican chef makes *ceviche* by soaking raw seafood in limejuice. The molecular basis for adaptation to extreme environments has been studied more for high temperatures than for any other extreme condition, but, even for these well-studied conditions, no straightforward and universally applicable rules have been uncovered that define whether a protein will remain folded and active. One interesting observation has come to light, though: enzymes derived from thermophilic organisms are as stable at thermophilic temperatures as our enzymes are at our body temperature. For example, testing each enzyme at the optimal growth temperature of its source organism yields similar values for key parameters including thermodynamic stability and enzymatic activity. Constant stability and activity are achieved, however, via a large number of subtle molecular interactions, which combine to produce a complex overall picture of thermostabilization.

Cold Adaptation

While hot springs and hydrothermal vents are important for both biology and geology, most of the water on our planet is rather cold. The bulk

*That these thermophiles seem to be the closest existing relatives to LUCA has widely been taken to suggest that life arose under high-temperature conditions, perhaps at the deep-sea vents. But remember: LUCA was quite advanced and a great many organisms had evolved and gone extinct long before she came on the scene. An alternative scenario is that life on Earth arose someplace far removed from the vents and evolved into organisms that filled many niches (including thermophiles living in the vents), only to have all of the nonvent organisms become extinct. How might this have happened? A meteorite or comet impact large enough to boil most of the volume of the oceans would nicely kill off everything but the deep-ocean thermophiles.

of the deep sea is at a constant temperature of 2°C, while in the vicinity of the polar ice caps, liquid seawater may even be cooled to below 0°C, as the typical salt content of seawater (3.4%) lowers the freezing point to −1.8°C. When seawater freezes, it is pure water that crystallizes, and this crystallization increases the salinity of the remaining liquid up to 15%. Under these conditions the freezing point may be depressed to as low as −12°C. Numerous species, collectively known as psychrophiles, are now known to thrive even under these seemingly harsh conditions.

The challenges associated with cold adaptation are very different from those associated with a thermophilic lifestyle. Whereas high temperatures speed up reactions—including those that break down biological macromolecules into smaller and much more stable molecules, such as carbon dioxide and water—low temperatures slow down reactions and thus merely bring biochemical systems to a halt rather than destroying them. The real danger sets in when water is allowed to freeze, as its large expansion during freezing exerts shear forces that can easily cause mechanical damage to cells and their components.

Two broad solutions to the problem of ice-induced damage have evolved on Earth. The first is the production of antifreeze compounds to prevent the formation of ice. Different species have come up with different ways to avoid freezing. Some organisms simply produce large quantities of osmolytes, such as glycerol, which lower the freezing point of their tissues by several degrees. Fish of the polar waters have evolved several different kinds of antifreeze proteins, ranging from the simple type I AFP (for Anti-Freeze Protein) that consists only of a single α helix, through to complex, high-molecular-weight glycoproteins. These proteins typically recognize and bind to structural features of nascent or growing ice crystals, thus blocking their further growth. As an alternative solution, some species of frogs and turtles have gone in the opposite direction. They have evolved ice nucleation proteins, which actually facilitate freezing of the animal's body liquids. By triggering ice formation in many different places at once, these proteins ensure that crystals cannot grow large enough to cause mechanical damage. Essentially, the freezing process is similar to what happens when a small sample of biological material is thrown into liquid nitrogen (at −196°C: the water freezes instantly in all places at the same time, which minimizes the damage to cell structures).

Ice-induced damage, however, is not the only problem that cold-adapted organisms face. Many of the molecular interactions that define our cells are affected by cold, and thus many aspects of biochemistry must be modified for an organism to thrive at low temperatures. These adaptations include the modification of membrane lipids (to avoid their stiffening, just as oils often solidify in the refrigerator) and the production of proteins that stop RNA molecules from binding to themselves to form complex structures that could inhibit translation.

A particularly successful strategy of adaptation to the coldest climates on Earth involves the collaboration between two species. Lichens, which colonize much of the rock surfaces in Antarctica, are famously resistant to the dry, cold conditions of their habitat. Although many lichens look like plants, they represent a symbiotic life form made up of a fungus and a photosynthetic organism. The ability to form lichens is spread widely across many families of fungi. In fact, it is estimated that one in five fungal species can perform the trick. Their symbiotic, photosynthetic partners can be algae (e.g., *Prasiola crispa*) or cyanobacteria (e.g., *Nostoc commune*), representing the domains of Eukarya and Bacteria, respectively. The symbionts make a very hardy team. Whereas *P. crispa* can survive on its own in the wet and relatively mild climates found on the coast of Antarctica, under the drastic temperature swings typical for the dry rock faces, its ability to survive as a lichen is unrivalled. Cold-resistant lichens, forming characteristically colorful plant-like structures, thrive at very high polar latitudes—even, for example, 2,500 meters up in the Horlick Mountains, which, at 86° south, are within 400 kilometers of the pole. Amazingly, lichens can typically withstand temperatures that range from −196°C to +30°C without noticeable effect.

An alternative to molecular adaptations that provide protection against freezing is provided by environmental protections against freezing, as is seen in the ice on and around Antarctica. Algae, for example, have been shown to survive in tiny liquid pockets encapsulated in the sea ice that surrounds Antarctica during the winter. As the freezing of a closed volume of seawater brings with it an increase in salt concentration, the freezing point is depressed in these environments. This prevents the organisms from freezing, but as a consequence they have had to adapt to high salinity as well as low temperatures and low light. Other species "hibernate" for most of the year in the sea ice and spring to life only when their habitat is defrosted for a brief period during the southern summer. Similarly, several species of algae have been found to grow a few millimeters under the surface of translucent rocks in Antarctica's "dry valleys," harsh, cold, and extremely dry areas that are widely regarded as the closest thing to Mars on Earth. The overlying rock acts like a greenhouse, significantly extending the growth season of these "endolithic" organisms and protecting them from drying out.

A final "ice niche" is located far below the surface of the ice cap in permanently liquid lakes thought to have existed for millions of years.

Kept warm by geothermal energy (helped by the insulating properties of a kilometers-thick ice cap), these lakes are thought to have been isolated from the rest of the biosphere for the entirety of their existence. Although they were discovered only in the 1990s, during ice-penetrating radar studies, these under-ice lakes are vast. The largest one, Lake Vostok, is located at 77° south beneath the eponymous Russian research station and is thought to contain as much water as Lake Ontario. In 1998, an international research team based at the Vostok station drilled more than 3 kilometers into the ice, deliberately halting about 100 meters above the expected lake surface. From the lower parts of the drill core, which contained ice formed from lake water, the researchers recovered evidence of microbial life. This finding strengthened the scientists' belief that the lake itself may well contain a unique ecosystem. Any exploration of the lake water itself would therefore have to proceed with technologies that are certain not to damage this biotope by contamination from the outside world. Moreover, a simple hole drilled into the lake would release the high pressure normally exerted on the lake water, thereby diminishing its freezing point depression. This might result in a loss of much of the lake water to freezing. Therefore, alternative drilling methods, possibly involving a drill robot that closes the tunnel behind itself, are being explored.

At the time of writing, there is disagreement between research teams from different countries on the best strategy to follow to avoid contaminating these pristine ecosystems with surface organisms. Depending on the outcome, the drilling into Lake Vostok may or may not resume within the next few years. The motivation for such studies, though, is clear. The Antarctic lakes, and the methods developed to explore them, are of interest as potential models for the saltwater oceans suspected to be hiding under the ice crusts of several moons in the outer Solar System, which we discuss in detail in chapter 10. Survival in these cold, subice lakes may also be relevant for the understanding of life's history on our own planet, which is believed to have undergone several "snowball Earth" episodes as recently as 600 million years ago, when the carbondioxide cycle (chapter 3) that regulates the Earth's temperature may (or may not—this is still controversial!) have gone haywire.

Drought and Salinity

As we have seen, many extremophiles can survive and even thrive at temperatures above 100°C or below 0°C, as long as water remains liquid (because of high pressure and/or salinity). Generally, the availability of liquid water is widely seen as a key requirement for life. Deserts and salt lakes illustrate the struggle for survival in environments where water is either absent or unavailable for chemical reasons.

All organisms we know of need water as a solvent for their biochemical reactions. But some have evolved ways of surviving long periods of drought in a passive state, and then carry on active living when the water returns. The bacterium *Deinococcus radiodurans*, for instance, attracted scientists' attention with its extraordinary resistance to ionizing radiation. First discovered in a can of corned beef that had been "sterilized" with γ-rays, it was also isolated from cooling baths of nuclear reactors. Research showed that the bug hosts a highly efficient DNA repair mechanism, which enables it to survive radiation levels a thousand times higher than would kill a human. As it is clear that radioactivity has not been the selective factor that produced this mechanism (nuclear reactors being a relatively new habitat), the question as to why a bacterium would have developed this trait was at first a mystery. The consensus now, however, is that the *D. radiodurans* DNA repair mechanism evolved as a means of surviving DNA damage induced by severe drought, as both irradiated and dried cells suffer similar types of DNA strand breakage due to the presence of active chemical species such as oxygen radicals.

Another impressive defense against the threat of drying is that employed by the tardigrades—a group of microscopically small animals found on all continents, and more endearingly known as "water bears." Tardigrades are typically found in water droplets suspended in moss and lichens. Remarkably, these organisms have two separate, highly original emergency routines. One is for the case of flooding and associated oxygen shortage, and it involves inflating to a balloon-like state that floats at the surface of the water. The second emergency plan is for the opposite case. When their habitat dries out, tardigrades shrink into a spore-like granule known as the tun state (*tun* is an archaic word for "barrel," which is what it looks like). In the tun state, the tardigrade replaces most of its water with the sugar trehalose. Trehalose solutions, unlike the solutions of most sugars, form an amorphous glass rather than sharp, damaging crystals when they evaporate down, and thus trehalose is an ideal medium for dehydrating proteins and DNA without damaging them.

Once the tardigrade converts into a tun, it is one of the toughest animals on Earth. Tuns found in moss samples from museums have been revived after more than a century of drought. Among other improbable stress treatments, researchers have suspended tuns in perfluorocarbon solvents and subjected them to hydrostatic pressures of up to 6,000 atmospheres, which more than 80% of the animals survived. Tuns can similarly survive temperatures as high as 150°C and as low as 0.2°C above absolute zero. If anybody wanted to send animals to Mars (and who doesn't?), tardigrades would probably be the most likely to survive the trip without much in the way of life support.

In addition to drought-stricken landscapes, there are ecosystems on Earth where water is present but not readily available. Think of the so-called Dead Sea, which despite its name is very much alive. Evaporation lakes like the Dead Sea are saturated with salt, so all the water is essentially taken up with the task of solvating salt ions. Living organisms have to compete with the ions for water, lest the osmotic gradient suck them dry. Several species of algae, along with bacteria and archaea, have adapted to high-salt environments such as that imposed by the Dead Sea. While the algae tend to be halotolerant (i.e., they can live with the salt but may be better off without it), many of the archaea found in salt lakes are obligate halophiles (i.e., they grow only at high salt concentrations). In response to the salt, all these adapted organisms maintain very high concentrations of other solutes in their cytoplasm, to keep their insides in osmotic balance with the outside world. While salt-adapted algae and bacteria tend to use small organic molecules for this purpose, highly halophilic archaea fight fire with fire by keeping extremely high concentrations of potassium chloride in their cells. This literally takes away the pressure from the membrane, but it shifts the stress and the requirement for adaptation onto the molecular machinery of the cell. All the proteins in a halophile have to be optimally folded and functional under saturated salt conditions, in much the same way that the proteins of hyperthermophiles function near 100°C. Researchers have therefore studied the amino acid sequences, structures, and functional characteristics of halophilic proteins in comparison with thermophilic and mesophilic ("normal") proteins in order to gain some insights into the evolutionary strategies used to adapt proteins to stress conditions, but the picture remains far from complete.

Extremes of pH

Another part of the "normality" that we rarely question is that the pH of most liquids in a biological context is close to the neutral value of 7, or possibly slightly above that (fig. 8.3). Where acids come into play, as in saliva and gastric juice, they are meant to destroy biological material and any surviving food-borne organisms. However, there are habitats (such as the Yellowstone solfatare fields described above) where both fungi and archaea thrive at pH values around zero, which is even more acidic than pure gastric juice. Although extremely acidophilic organisms have been

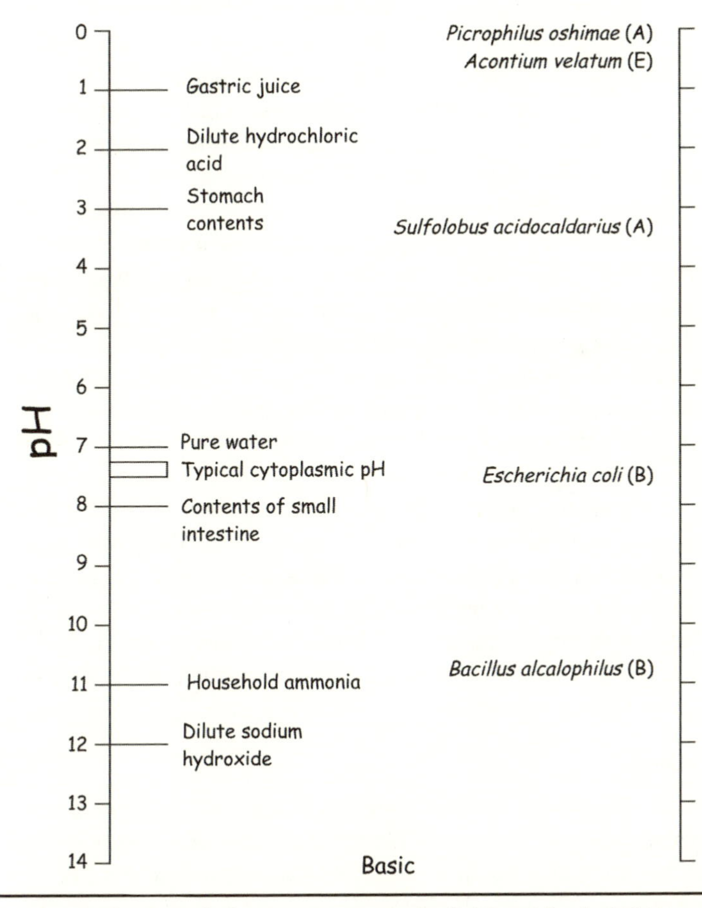

FIG. 8.3. The pH scale for some representative fluids and for the habitats of some pH extremophiles (A, archaeon; B, bacterium; E, eukaryote). Energy production in most aerobic organisms requires their external pH to be lower then their internal pH. Thus life extends farther toward the acidic end of the range than toward the basic end.

known for decades, the nature of their specific adaptation is far less well studied than that of thermophiles.

The base-loving alkaliphiles, by contrast, have been studied in some detail by researchers interested in bioenergetics and the electrochemistry of the living cell. The interest was motivated by the special challenge these cells face. Most cells make use of the fact that their internal pH is higher (more basic) than the pH of the medium outside. For instance, the enzyme that synthesizes most of the energy-carrying ATP produced by aerobic metabolism (ATP synthase, or F_1F_o ATPase) requires movement of hydrogen ions in a millstream-like flow,* from the acidic outside of the cell to the basic inside, to drive the production of energy-rich ATP from ADP and inorganic phosphate. If the pH gradient is reversed, the enzyme can work the other way round, consuming ATP reserves and pumping protons out of the cell. Active transport of many other kinds of ions and a number of nutrients also relies on the pH gradient. Thus the adaptation to high pH involves some drastic reorganization of metabolic electrochemistry and, perhaps for this reason, is found in only a few branches of the tree of life. The standard model organism for alkaliphile studies, *Bacillus alcalophilus,* can thrive at pH values well over 10, while maintaining its cytoplasm at a fairly normal pH of 8.6.

For a microbe living in challenging physical conditions, such as in 100°C hot springs or deep beneath the surface at pressures of hundreds of atmospheres, there is no way the cell can wall itself off from the challenge. Under such conditions, the stress affects the cell's entire molecular machinery, all of which must be adapted in response. In contrast, defenses against extremes of pH and other specific chemical challenges, such as high levels of toxic metal ions, are generally massed at the membrane, where the stress can be held at bay. In terms of evolution, this suggests that adaptations to challenging chemical conditions are "cheaper" —require the alteration of fewer cellular systems—than adaptations to challenging physical conditions.

Going Deep

The effects of pressure on organisms are not so widely appreciated. Changes in atmospheric pressure may be indicative of a change in the weather, but otherwise they don't normally affect organisms that, like us, live on dry land. When we climb a mountain, the decline in the partial pressure of oxygen affects us quite strongly, but the overall change of atmospheric pressure will not have any notable effects.

The situation changes drastically when we start diving into the ocean's depths (fig. 8.4). As a rule of thumb, hydrostatic pressure increases by 1 atmosphere for every 10 meters we descend beneath the surface. The increasing pressure and the associated changes in the solubilities of gases and toxicity of oxygen limit the reaches of recreational scuba diving to around 40 meters, and those of divers with optimized gas mixtures to around 500 meters. Beyond that, humans need submarines to explore life under hydrostatic pressure, and that is one of the reasons we know

*The analogy is better than you might imagine. In 1997, Masasuke Yoshida and his research team filmed the F_1F_oATPase rotating as it synthesized ATP.

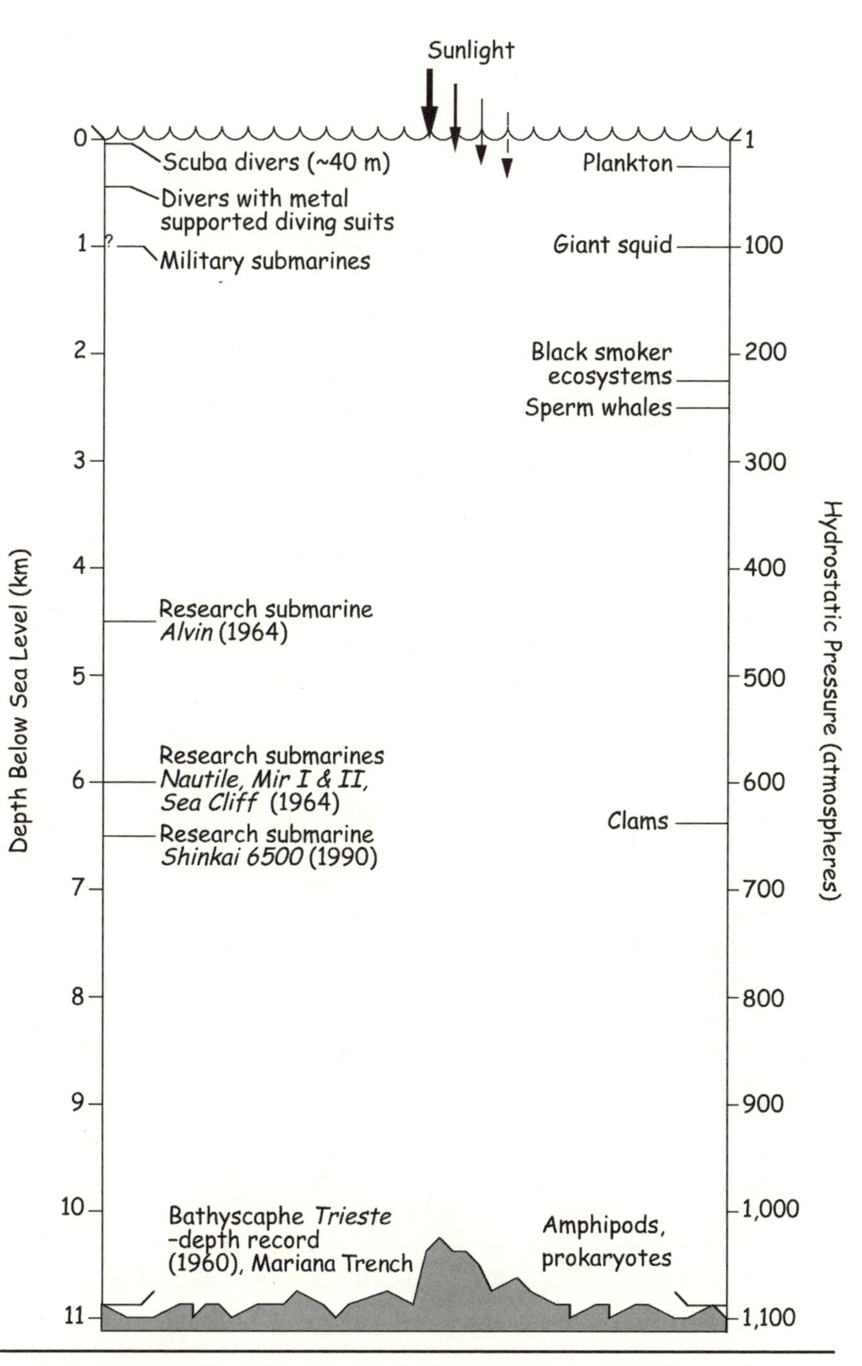

FIG. 8.4. Pressure adaptations allow barophilic organisms to thrive even at the deepest depths of the ocean.

a lot less about barophiles (pressure-adapted organisms) than about thermophiles.

There is no doubt that barophilic adaptation exists and is widespread in the oceans. Research vessels have consistently found life on the seafloor over the whole depth range down to the 11-kilometer maximal depth of the Mariana Trench. This implies that there are living organisms adapted to pressures of up to 1,100 atmospheres. Most ordinary microorganisms, in contrast, stop growing well below 500 atmospheres. Mammals like us, with lungs and an entire metabolism depending critically on the equilibrium of gases (oxygen and carbon dioxide) transported through the bloodstream, are intrinsically pressure sensitive. Among mammals, the sperm whale with its 2,440-meter proven diving range is a very lonely record holder.

The difficulty in obtaining samples from deep-sea habitats, along with the challenges of conducting biochemical experiments under high pressure conditions in the laboratory, have conspired to make this research field one of the less comprehensively studied. It was, for example, only a few years ago that researchers identified the first genes unequivocally involved in adaptation to high hydrostatic pressure. Most of the mechanisms that allow life to thrive even at the greatest depths of the oceans remain to be discovered.

High-pressure ecosystems are not limited to the seafloor. Organisms in the subglacial lakes of the Antarctic are also exposed to extremely high pressures. Similarly, subsurface habitats deep within the Earth's crust must be exposed to crushing pressures, combined with high temperatures and extremely limited resources. Nevertheless, recent evidence suggests that these extreme environments, too, may be teeming with life. In fact, the deep subsurface is one of the areas where the limits of the biosphere are far from clear. In isolated cases, microbes have been recovered from subsurface habitats as far as 1,500 meters below ground. Like the deep sea, the deep subsurface is cut off from sunlight and the food web based on photosynthesis. Organisms living there have to find alternative sources of energy in the minerals around them.

A spectacular example of a subsurface habitat is provided by Movile Cave in Romania. In 1986, a complex subsurface ecosystem was discovered in the cave, and follow-on studies in the 1990s showed that bacterial mats degrade the limestone of the cave walls and thereby supply the carbon for an entire, relatively isolated ecosystem made up of around fifty interdependent species. It is thought that the energy required to drive the ecosystem is derived from the oxidation of sulfides that enter the cave via the groundwater. Isolated underground ecosystems such as Movile are of great interest to astrobiology, as life on other planets may well have survived in such niches after the surface became sterile due to the loss of the atmosphere or liquid water. Note, however, that they are not exact models of independent (extraterrestrial) ecosystems, as oxygen is, of course, the product of photosynthesis.

In contrast to Movile Cave, other subsurface ecosystems may be truly independent of sunlight as an energy source. The idea of such deep, hot biospheres was tirelessly promoted by the Cornell University astrophysicist Thomas Gold (1920–2004), but has only recently gained anything like widespread acceptance. Gold's ideas regarding life below the Earth's surface were an extension of his somewhat heretical theories about the origins of fossil fuels. The traditional view is that coal and oil are the remains of long-dead plant matter and marine algae, respectively. Gold believed, instead, that these hydrocarbons were incorporated into the Earth during accretion and provide a ready carbon source for a complex and rich underground ecosystem that could even exceed the mass of the surface biosphere we see around us.

And is there any evidence for Gold's claims? With regard to the idea of at least a limited underground biosphere the answer is increasingly moving toward yes. For example, in 1990, a hole drilled some 6.7 kilometers into the granite of Siijan, Sweden, turned up a thick, black, strongly odiferous fluid from which two different species of thermophilic, iron-reducing bacteria were later cultured. About the same time, complex ecosystems were being reported within rocks freshly harvested from 3 kilometers down in South African gold mines. But the idea of microbes living in what seemed to be solid rock far beneath the surface didn't sit well with the microbiology community. In fact, it wasn't until 1993, when Todd Stevens and James McKinley of the Pacific Northwest Laboratory probed for ecosystems deep below eastern Washington state, that microbiologists really stood up and took note. McKinley and Stevens identified methanogenic microbes from basalt collected aseptically from several hundred meters down in a thick bed of basalt. Methanogens use the energy available from the conversion of carbon dioxide and hydrogen to methane and water. In their better-known habitats (such as cow stomachs), the hydrogen is produced biologically, but the water samples from the deep drill contained more hydrogen than could be explained by any biological mechanism. Investigating further, the researchers found that the observed amount of hydrogen can be produced by the reaction of reduced iron from the basalt with hot, oxygen-free groundwater. Putting these observations together, the researchers theorized that their subsurface ecosystem is fueled by abiologically produced hydrogen and carbon dioxide, and is independent of solar energy in any of its many guises. To clinch their hypothesis, the researchers demon-

SIDEBAR 8.3

Weighing the Probabilities

The discovery that organisms could live in seemingly solid rock deep beneath the Earth's subsurface opened the door to a great deal of speculation about how life might exist in similar habitats throughout the galaxy. This was particularly true given that the subsurface ecosystem found in eastern Washington state survives without recourse to solar energy in any of its many guises; every other ecosystem on Earth either requires light for photosynthesis or uses oxygen, a byproduct of photosynthesis. Since, for example, Mars may be geologically active enough to support liquid water within its crust (more on this in the next chapter), the discovery of rock-eating organisms on Earth seems to suggest that similar locales could be inhabited on Mars. But is it that simple?

Probably not. The problem is that, even if a planet contains environments that could support life today, it is still very much a question as to whether life could have arisen there originally. As we saw in chapter 5, special conditions might be required for the spark of life to arise from inanimate matter. Thus the enormous range of bizarre environments that life can inhabit on Earth probably reflects life's extraordinary ability to diversify to fill any available niche, rather than the ease with which life can arise in the first place.

strated that their bacteria could survive for more than a year when sealed in a flask with nothing more than hot water, carbon dioxide, and basalt to munch on. Indeed, the bugs not only survived, but actually thrived. Their numbers increased under these seemingly extreme culture conditions, suggesting that they are well adapted to their diet of reduced rocks, water, and carbon dioxide. Could similar ecosystems survive elsewhere in the Solar System? We explore this subject in depth in chapters 9 and 10 (but see sidebar 8.3).

Conclusions

In some sense the Terrestrial extremophiles simultaneously tell us both a great deal and very little about the ability of life to arise and thrive on other planets. On the one hand, the ability of life to survive in such unlikely, resource-poor environments as solid rock kilometers below the surface of Washington state suggests that the range of niches suitable for life may be much broader than many scientists previously suspected. Similarly, tardigrades' ability to survive desiccation to a water content of 3% suggests that oceans and rivers may not be required even for advanced, multicellular organisms. It is quite plausible that the rock-eating archaea that swarm within the Earth's crust could survive within the crust of Mars as well; and, although probably less likely, the sub-ice microbes of Lake Vostok could survive in the under-ice ocean of Europa. But even if such life could survive on Mars or Europa, could it have arisen there in the first place? Or was there something special about the early Earth that rendered it much more suitable than elsewhere in the Solar System for the formation of life? Perhaps the answers to these questions will be found as humanity reaches out to explore the cosmos—the topic of the final chapters of this book.

Further Reading

Deep-sea biotopes. Edmond, J. M., and Von Damm, K. L. "Hot springs on the ocean floor." *Scientific American* 248, no. 4 (1983): 78–93.

Extremophiles. Gross, Michael. *Life on the Edge.* Cambridge, MA: Perseus, 2001.

Strain 121. Kashefi, K., and Lovley, D. R. "Extending the upper temperature limit for life." *Science* 301 (2003): 934.

Subsurface ecosystems. Stevens, Todd O., and McKinley, James P. "Lithoautotrophic microbial ecosystems in deep basalt aquifers." *Science* 270 (1995): 450–54; Gold, Thomas. *The Deep Hot Biosphere.* New York: Copernicus Books, 1998.

CHAPTER 9

Habitable Worlds in the Solar System and Beyond

On the summer night of July 16, 1969, Neil Armstrong, Buzz Aldrin, and Michael Collins blasted off from Kennedy Space Center in Florida for humanity's first visit to the surface of another celestial body. Three days later the trio dropped into Lunar orbit, from whence Armstrong and Aldrin split off from Collins and took the lunar module *Eagle* down to the surface. On July 20, uttering his historic—but misspoken—phrase, "That's one small step for [a] man, one giant leap for mankind," Armstrong stepped off the lander and walked on the face of our Moon. After a short two and a half hours exploring the surface, the duo returned to the lunar module, spent a fitful night (during which Aldrin, a Catholic, performed the first extraterrestrial communion), and launched back to a rendezvous with the orbiting command module. Firing up the command module's single rocket, the reunited astronauts spent three more days on the return home, splashing down in the Pacific on July 24.

Upon their return, the three astronauts were immediately ushered into a fully self-contained and hermetically sealed isolation trailer (a converted recreational vehicle!), physically cut off from the rest of the world. Heroes they may have been, but they would remain behind tightly sealed doors and windows as if they, themselves, were some exotic Lunar samples. The reason? A fear that "Lunar bugs" that might have infected them could escape to wreak havoc on our planet.

After three days the trailer arrived in Houston, Texas, where the three astronauts joined three others (a doctor, a NASA public affairs officer, and a film technician who had accidentally been exposed to some Lunar dust while handling a film canister) in more spacious quarantine facilities at the Lunar Receiving Laboratory. While the doctor kept close watch on the health of the astronauts, scientists in other, equally well-isolated laboratories in the complex incubated Lunar dust in nutrient broths and injected it into mice to see whether any microorganisms could be cultured. After two and a half weeks, neither the humans nor the mice seemed any the worse for their exposure to the Lunar materials and the quarantine was lifted. The astronauts were released to a world tour, months on the banquet circuit, and changed lives.

In December 1969, some six months after *Apollo 11* achieved President Kennedy's goal "of landing a man on the moon and returning him safely to the Earth," *Apollo 12* set off for the Moon's prosaically—if inaccurately—named Oceanus Procellarum, Ocean of Storms. One of the many goals of this mission was to demonstrate that the *Apollo* technology could achieve a precision landing (*Apollo 11* had missed its target by several kilometers), so that future missions could explore more interesting, but difficult, terrain. The all-Navy crew (picked, in part, because of their expertise in navigating) pulled it off and landed within 200 meters of their target, the unmanned *Surveyor 3* spacecraft that had landed on the Moon two and a half years earlier. The two astronauts, Pete Conrad (1930–99) and Alan Bean, spent almost eight hours exploring the Lunar surface, during which they collected 34 kilograms of rocks and soil.* They also snipped off several pieces of *Surveyor 3* for return to Earth, so that engineers back in Houston could see how the various materials had fared after so many months under the harsh Lunar conditions. And what happened when the *Apollo 12* astronauts returned? They too were ushered into quarantine. Once again, though, no Lunar life was found—save, perhaps, some Terrestrial bacterial spores that may (or may not; there is some debate as to whether they were picked up after the material returned to Earth) have survived in hibernation on the Lunar surface, buried under some insulation deep within the *Surveyor* camera.

The lack of Lunar life was, in reality, to be expected. The argument against life on the Moon was and remains, basically, that if you wanted to build a really good sterilizer you'd make something like the Moon's surface: no atmosphere, no water, extremes of heat well past the boiling point, intense radiation, and intense ultraviolet light. Still, even with the insights provided by the intervening three decades, these precautions seem prudent. After all, what do we really know about the range of conditions that life in its broadest scope might find suitable? How does the broad range of environmental conditions to which life has adapted on Earth correspond to the conditions that exist elsewhere in the Solar System and beyond (again, given the caveat that the set of environmental niches that life can evolve to fill may be much larger than the set of conditions under which it can arise in the first place)? A key element of the relationship of life to the Universe is where, other than the Earth, life

*Conrad, ever the joker, snuck an automatic timer for his camera into a rock sample bag so that he could perch the camera on *Surveyor* and snap a picture of himself and Bean together. His hope was that, after the film was developed, people would look at the photograph and ask, with shock, "If both astronauts are in the picture, who took the picture?" Alas, though, when the time came Conrad could not find the timer in the sample bag where he'd hidden it.

might be harbored. Here we explore this issue in detail for the Solar System and, in necessarily less detail, for the rest of the cosmos.

Potential Abodes of Life Elsewhere in the Solar System

Let us start the search in the immediate neighborhood. Although the first astronauts to return after a Moon landing were placed under quarantine for fear of infection with Lunar life forms, it is now clear that our satellite never had the rich mix of volatile elements that are almost certainly required for life. If there are any microbes on the Moon, it is safe to assume that they traveled there with the astronauts and have been dead for several decades now.

Moving inward from our home planet we have Venus and then tiny, rocky Mercury. As described earlier (chapter 3), because Venus is closer to the Sun it lacks the cold traps required to keep water in its lower atmosphere, and thus Venus's water was, geologically speaking, lost to space rather quickly via photolysis. With the runaway greenhouse effect that then ensued on Venus (no water means all of the carbon dioxide stays in the atmosphere rather than, as on Earth, safely locked in carbonate rock), the temperature of the surface rose well above the melting point of lead. Precisely how long this took is still a matter of speculation. But the consequences for the possibility of life on our nearest planetary neighbor are clear: we don't know of any chemistry that is of sufficient complexity to support life that would be able to withstand Venusian conditions. And while there have been suggestions of cooler, more favorable conditions in the Venusian clouds, we should probably conclude that the overall probability of life on Venus is very low. Mercury, as hot as an oven and almost entirely lacking in atmosphere, seems an even more unlikely habitat. If there are any other denizens of the Solar System, it seems likely that they live out beyond the Earth's orbit.

Mars: From Canals to Rovers

From early on, it was clear that Mars deserves serious consideration as a potentially habitable planet. As it is the second closest planet to the Earth, even the modest telescopes directed at Mars in the late eighteenth century were sufficient to identify clouds, polar ice caps, and the length of its day: 24 hours and 37 minutes. More tantalizing still, by the late nineteenth century astronomers had noted a "wave of darkening," a seasonal color variation occurring on a global scale. Similar color variations are seen on Earth in, for example, New England or Scandinavia in the autumn. Could the Martian color changes also represent seasonal variations in plant coverage?

Attempts to detect life on the Red Planet have a somewhat checkered history, dating back to 1877. In that year the Italian astronomer Giovanni Schiaparelli (1835–1910) telescopically spied extended networks of trenches that he termed *canali,* the rather innocuous Italian word for "channels." Even though Schiaparelli never meant to imply that these structures had been constructed by intelligent beings, others—in particular the independently wealthy and exceptionally self-promoting American astronomer Percival Lowell (1855–1916)—eagerly translated *canali* as "canals." That is, as artificial structures. Lowell built himself one of the finest observatories of the era in the mountains of northern Arizona and spent decades visually mapping out the features on Mars and concocting a detailed hypothesis about a dying, intelligent race that had built the canals to carry water from the poles as their planet fell into drought. Not surprisingly, Mars quickly became a popular location of extraterrestrial life forms in the imagination of science fiction writers. The invaders trying to colonize Earth in H. G. Wells's classic *War of the Worlds* (1898) were only a vanguard of the many different civilizations placed on Mars by writers throughout the twentieth century. With his *Martian Chronicles* (1950), Ray Bradbury invented an extinct Martian civilization of which only the monuments remain. The possibility—nay, the certainty—of life on Mars was part and parcel of nineteenth- and early-twentieth-century thinking about our next neighbor away from the Sun.

Sadly, with the advent of improved telescopes and, perhaps not coincidentally, photographic methods that replaced visual observations, the case for intelligent life on Mars was quietly dropped; Lowell's linear canals have never been captured on film (fig. 9.1). In this enlightened—if less romantic—age, we know they were the creation of an overactive imagination, poor "seeing," and no doubt a sincere belief that we are not alone. Still, the same telescopic observations that killed the canals supported the claimed widespread seasonal color changes on Mars. Even as late as 1950, most authoritative sources believed this to be strong evidence for at least vegetative life. Cracks were apparent in this hypothesis too, however, as far back as 1909 when William Campbell (1862-1938) climbed above half of the Earth's atmosphere to the top of Mount Whitney (at 4,417 m, the highest point in the continental United States) to make the first spectroscopic studies of the Martian atmosphere. His studies indicated that Mars lacked observable oxygen or water vapor, casting doubt on Lowell's vision. Follow-on studies over the next five decades confirmed that Mars's atmosphere is both extremely dry and thin—boding poorly for the planet's climate—and consists mostly of carbon dioxide. There are no oxygen-producing green fields on the Red Planet.

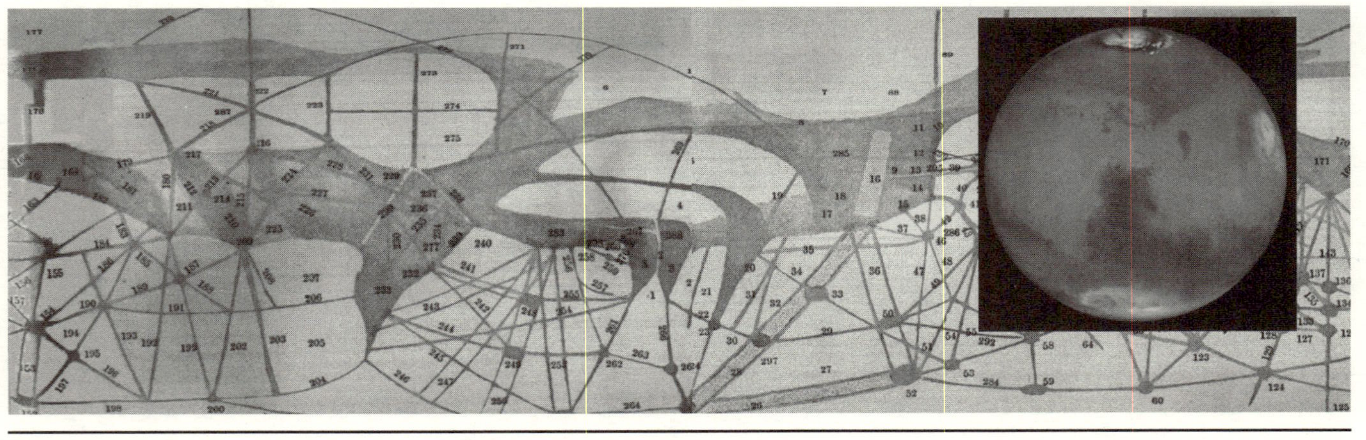

FIG. 9.1. Percival Lowell mapped the canals of Mars in exquisite, hand-drawn detail. Sadly, though, Lowell's linear canals did not survive the advent of astrophotography. Modern film or electronic images of Mars (shown in the inset is an image from the Hubble Space Telescope) fail to show any linear features. (Courtesy NASA/STScI)

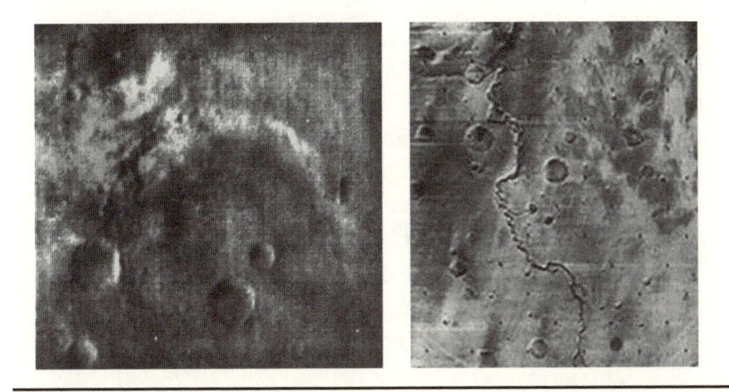

FIG. 9.2. *Mariner 4,* on the first successful Mars flyby, sent back pictures of a bleak, geologically dead landscape reminiscent of the Moon (left). The more global picture available from orbit, first seen by *Mariner 9,* is quite different (right): Mars's northern hemisphere is filled with the remnants of once active geology, including, perhaps, water-carved features such as the putative river channel pictured here. (Photos courtesy NASA/JPL)

The final nail in the coffin of Mars's romantic, life-filled image came with the first close-up investigations of the planet. In 1962, after one previous failed American and several failed Soviet attempts, the U.S. spacecraft *Mariner 4* flew 9,800 kilometers above the southern hemisphere of Mars, snapping twenty-two pictures along the way. Digitized and radioed back to Earth (at a trickling 8 bits per second), the photos supplanted centuries of speculation and revealed for the first time details of the surface of the Red Planet. What the pictures revealed was an ancient, heavily cratered landscape reminiscent of the Moon (fig. 9.2, left). Far from the dynamic, life-filled world of lore, the southern hemisphere of Mars appeared geologically dead, without even the dynamism to erode away the many craters dotting its surface. Follow-on flybys by *Mariner 6* and *Mariner 7* in 1969, also of the southern hemisphere, returned another two hundred close-up pictures of ancient, heavily cratered terrain. Not a canal in sight.

The generally pessimistic funk that Mars research fell into after the flyby missions did not lift until 1971, when *Mariner 9* became the first spacecraft to orbit a planet other than Earth (while its partner, *Mariner 8,* was resting on the bottom of the Atlantic Ocean due to a launch failure). As the craft approached, the Martian disk appeared strangely featureless; the planet was engulfed in a dust storm of global proportions. Although it barred the view to the planet's surface, the storm did explain one Martian mystery: the seasonal color changes long thought to indicate changes in vegetation reflect instead seasonal changes in the distribution of dust. When the dust cleared a few weeks later, however, the outlook was brighter than expected. *Mariner 9* found evidence of a far more active planet than dreamed of after the earlier flybys; only the *southern* hemisphere of Mars is old and cratered. Much of the remainder of the planet shows abundant signs of past geological activity, including extinct volcanoes up to twice as high as Mount Everest and a canyon system that would dwarf our Grand Canyon.

Of particular interest to us, some of the exciting geology that *Mariner 9* turned up seems to have been created by a fluid, perhaps water (fig. 9.2, right). This evidence included what looked like vast sedimentary deposits in the polar regions and extensive networks of valleys that looked very much as if they were formed by rivers. But the orbiter also made detailed studies of the Martian weather and found that, while the Martian tropics can top out at a downright pleasant 24°C during the warmest summer days, the temperature on Mars is typically quite far below freezing, and the planet's atmospheric pressure is too low to allow water to exist as a liquid. Thus the post–*Mariner 9* picture was of a now frozen and exceptionally dry planet with only ancient relics of a presumably wet and watery past.

Still, optimism about Mars ran fairly high during the 1970s, when follow-on studies from the *Viking 1* and *Viking 2* orbiters (we discuss the associated *Viking* landers in the next chapter) carried much better cameras into Martian orbit. With these improved cameras, scientists back on Earth cataloged many apparently fluvial (water-formed) features, including presumably water-carved, teardrop-shaped "islands" within massive outflow channels that themselves seem to have been carved by intensive, if brief, floods, perhaps exceeding by a factor of 10,000 the flow rate of the mighty Mississippi River. As further evidence of these massive floods, the orbiters spied vast areas of chaotic terrains reminiscent of the scablands of the Pacific Northwest that were generated at the end of the last ice age, when an ice dam broke and a volume of water the size of Lake Superior flooded large parts of what are now Idaho and Washington states. It was even thought that the *Viking* orbiters had spied fossil remnants of the shorelines of some ancient Martian ocean. Based largely on these images, the warm and wet early period of Martian history seemed assured. And if there was an ocean, isn't it likely there may have been life?

Unfortunately for those of us who would like to find Mars inhabited, the tide has once again largely turned against the warm, wet early Mars hypothesis. One of the most fundamental problems with this idea is that, as we discussed in chapter 3, the early Sun is thought to have been some 20% dimmer than it is today. And if sunlight cannot now heat Mars above the freezing point of water, how could it have done so billions of years ago? Various greenhouse gases have been proposed that could have bumped the temperature up, but to date it has been hard to come up with a physically plausible scenario by which Mars remained largely above freezing during the earliest days of the Solar System. Consistent with these arguments, the high-resolution infrared spectrometer on board the orbiting *Mars Odyssey* (launched in 2001, and named after the spaceship in Arthur C. Clarke's novel *2001: A Space Odyssey*) failed to pick up any of the spectral signatures of carbonate rock. Given that CO_2 forms carbonates, such as limestone, whenever liquid water is present on Earth, the lack of such carbonates suggests that Mars has never hosted significant bodies of standing water. It now seems generally accepted among planetary scientists that Mars never had a long warm, wet spell. But then where did Mars's fluvial erosion features come from? There's been no end of speculation about this. The possibilities that have been suggested range from wind erosion, to temporary running water—either after a meteor strike vaporized some ice and cranked up the atmospheric density or underneath thick, insulating ice caps—to sapping (erosion from within) caused by the eruption of geothermally heated groundwater (Mars features, after all, enormous and relatively fresh-looking volcanoes, and thus a liquid, subsurface aquifer is definitely within the realm of possibility). However they were formed, though, it is widely held that the fluvial features seen so vividly from orbit stem from exotic processes or ephemeral conditions and do not reflect a consistently warm and wet period in Martian history.

Still, while the idea of a long warm, wet period in Mars's early history has receded, we do have tangible evidence of liquid water having once flowed on the surface of the planet. NASA's *Opportunity* rover touched down (bounced to a stop, actually, on airbags) on January 25, 2004, on Mars's Meridiani Planum, a smooth, flat plain the size of the state of Colorado that had been selected because spectroscopic investigations from orbit indicated it was decorated with the mineral hematite. Hematite, an iron mineral, can be formed by several mechanisms, but on Earth it is most commonly deposited in aqueous environments; NASA was abiding by its Mars creed: "follow the water." Cushioned by its giant airbags, the craft bounced a dozen times on the Martian surface before finally coming to rest in a small crater. When the rover shook off its rough landing, stood up, and turned on its cameras, it found itself staring at thick layered beds (fig. 9.3). The *Opportunity* rover, the fifth successful Mars lander, was the first to find bedrock.

On Earth, the majority of layered rock is sedimentary, laid down over successive seasons or successive floods by liquid water. But successive

FIG. 9.3. The *Opportunity* rover (foreground), the fifth successful Mars lander, bounced down onto the plains of Meridiani and struck a cosmic hole-in-one by landing in a small crater dubbed Eagle. Turning on its cameras, it imaged the first bedrock ever identified on the Red Planet. Detailed imaging and spectroscopic investigations suggest that the crater wall consists of sedimentary rocks laid down from liquid water—briny, highly acidic water, but liquid water nonetheless. (Courtesy NASA/JPL)

volcanic eruptions (either as lava flows or as successive layers of ash) can also form layered rock, as can changing wind patterns. So which was it at Meridiani? The bedrock in *Opportunity*'s crater showed clear evidence of cross-bedding, layers formed at angles, as is often seen in stream beds on Earth due to turbulent flow. Some of the rock surfaces also showed clear signs of polygonal cracks, reminiscent of the hexagons that sometimes form in drying mud. The layers were also filled with small, round rocks that also littered the ground at Meridiani. Termed "blueberries" after the fruit they seemed to resemble in shape and size if not color (somehow "grayberries" just sounds wrong), the small spheres littering the crater are composed of hematite, again most likely formed in situ in the rocks by the action of water.

Perhaps even more revealing than the gross structures of the strata are their chemical compositions. The rover's α-particle and x-ray spectrometer (APXS), which can identify the elemental composition of rocks (see sidebar 9.1), indicated that the rocks at Meridiani contain large amounts of magnesium, calcium, and iron sulfates, along with traces of chlorine and bromine. On Earth, such salts form preferentially by aqueous deposition and are a near certain sign that liquid water was once present. Similarly, *Opportunity*'s Mossbauer spectrometer found evidence that some of the iron at Meridiani is tied up in the mineral jarosite, a hydrated form of potassium iron sulfate that, on Earth, is invariably formed by the aqueous leaching of iron minerals under acidic conditions. So it looks as if we have firm evidence that at least parts of Mars were wet sometime in the distant past (albeit with rather acidic brine), but that Mars then took a very different environmental turn.

So where is the water on Mars now? Some of it is tied up in the polar caps, which consist of a seasonal mix of water ice and frozen carbon dioxide (Mars is so cold that a significant fraction of its atmosphere condenses out at its poles each winter). But the polar ice caps do not account for much water: the larger north polar cap is estimated to contain 1,200,000 cubic kilometers of water ice, which is less than half the size of the Greenland ice cap and a mere 4% of the volume of Earth's Antarctic ice cap. To search for possible reservoirs of water outside the polar caps, the *Mars Odyssey* carried a neutron spectrometer. This instrument maps out the distribution of hydrogen within the first few tens of centimeters of the Martian surface (see sidebar 9.1), which, given the abundance of oxygen in planetary crusts, is almost certainly tied up in water. And what did the neutron spectrometer find? It found massive deposits of hydrogen in the soil at high latitudes, and progressively less as one approached the equator. In fact, the soil in wide swaths of the northern and southern latitudes of Mars seems to consist of at least 50% water by weight; Mars is covered in permafrost (fig. 9.4).

Exciting as the neutron spectrometer discovery was, permafrost is not the stuff upon which life is founded. But if there is water under the surface of Mars as ice, is there some as liquid as well? The European Space Agency's orbiting *Mars Express* has carefully mapped the upper reaches of several of the planet's larger volcanoes and found that they are so free of craters that the terrain must be less than a few million years old. While that's a long time for us, it's less than 0.1% of the age of the planet, suggesting that Mars is likely still geologically active today (after all, what are the chances that Mars was active for more than 99.9% of its

SIDEBAR 9.1

Sniffing out Habitats

So you're going to send a robot a hundred million miles away from home to investigate Mars for signs of water. What do you pack? In part that depends on what type of mission you are planning. Orbiters and landers face different opportunities and demand different, if often complementary, technologies.

Because orbiters, obviously, remain at a distance from the object to be investigated, they need to be equipped with *remote sensing devices.* A good camera, of course, is a mandatory accoutrement for any tourist, and the best cameras in Mars orbit now approach submeter resolution. But in addition to pictures, you want to be able to measure the spectral properties of the atmosphere and surface. That is, you want to see which wavelengths of visible, ultraviolet, and infrared light the surface absorbs. The UV wavelengths allow for the identification of atmospheric components, whereas at infrared wavelengths both atmospheric and surface components exhibit highly characteristic "fingerprints." The infrared spectrometer also provides a means of determining the temperature of the surface (calculated from the known temperature-emittance relationship of a black body) and, indirectly, its structure: loose soils cool rapidly after nightfall, whereas rocks hold their heat and thus cool more slowly.

The modern Mars orbiter also wouldn't be caught dead without γ-ray and neutron spectrometers. The γ-ray spectrometer can detect the relative abundances of a few radioactive elements from the γ-rays they emit when they decay. It can detect even more elements from the cosmic-ray-induced emission of γ-rays. When cosmic rays (mostly high-energy protons) strike the surface of an airless or near airless planet (the Earth's atmosphere, for example, would screen out both the cosmic rays and the induced γ-rays), they excite atomic nuclei in the soil. The excited nuclei quickly lose their extra energy by emitting γ-rays characteristic of their element type. The neutron spectrometer runs along similar lines. Some of the cosmic rays knock neutrons out of atomic nuclei in the soil. If these high-energy neutrons strike a heavy nucleus, they bounce off without losing much speed (imagine a Ping-Pong ball bouncing off a bowling ball). If, instead, they strike a low-mass nucleus like hydrogen, they lose much more of their kinetic energy (imagine a Ping-Pong ball bouncing off another Ping-Pong ball). By measuring the kinetic energy of neutrons that emerge from the Martian surface, the neutron spectrometer indirectly detects the amount of hydrogen (presumably as water ice) in the soil; if the measured neutrons are moving rapidly, the soil contains only heavy nuclei, whereas if the neutrons are moving slowly, the soil contains hydrogen. Together the neutron and γ-ray spectrometers can map out the abundances of a few dozen elements in the first few micrometers to tens of centimeters of Martian surface with 10- to 100-kilometer spatial resolution.

In contrast to cameras and spectrometers, which are effectively passive instruments, some remote sensing devices are active. The *Mars Global Surveyor,* for example, carries a laser altimeter that bounces an infrared laser pulse off the Martian surface in order to map out the planet's topography in stunning detail. And the *Mars Express* orbiter carries a sounding radar experiment that is on the lookout for subsurface water. These are heady days for Mars exploration!

In contrast to orbiters, rovers get up close and personal with the rocks and thus require a more "hands-on" suite of instruments. Imaging, of course, is still critical (if for no other reason than the folks back home who are paying for the trip want some nice color pictures to gaze at over breakfast), and for more detailed mineralogical analysis, imaging spectrometers (or cameras with lots of color filters) are essential. But we also want to monitor the mineral contents of rocks more directly. Two approaches have been adopted in recent rover missions. The first, the α-particle and x-ray spectrometer (APXS), determines the elemental composition of rocks and soils to complement the spectroscopic mineral analysis performed by the imaging system. The APXS works by exposing samples to energetic α-particles and x-rays from a small amount of radioactive curium, and then measuring the energy spectra of the α-particles and x-rays that are scattered back from the sample into the detector. The energy of the scattered α-particles is sensitive to the weight of the atomic nuclei they strike, with lighter nuclei producing larger changes in the energy of the α-particles (à la the Ping-Pong ball analogy described above). Thus the α-particle mode on an APXS spectrometer is particularly sensitive to lighter elements such as carbon and oxygen. The x-ray mode works via the excitation of x-ray fluorescence in a sample, and is particularly sensitive to important mineral-forming elements such as magnesium, aluminum, silicon, potassium, and calcium.

The second approach is Mossbauer spectroscopy. In contrast to the elemental analysis performed by the APXS, the Mossbauer spectrometer determines the *chemical* makeup of minerals containing one specific element: iron. At the heart of this spectrometer is a small chunk of radioactive cobalt-57. When this decays, the resulting iron-57 (^{57}Fe) is left in a high-energy state. When that, in turn, decays, it emits a γ-ray photon that can be absorbed by ^{57}Fe nuclei in the sample. The precise wavelength of the absorbed γ-radiation is sensitive to the chemical compound in which the iron resides, thus allowing Mossbauer spectroscopy to characterize the chemical nature of iron minerals that are otherwise difficult to detect. Both the APXS and Mossbauer spectrometers require physical contact with the sample and long integration times (typically overnight), and thus both are well suited to rover missions that can move the spectrometers from rock to rock.

Last but not least, if you're going all the way to, for example, the Mars surface, you might also want to toss in a RAT, or "rock abrasion tool." The issue is dust; the winds of Mars stir up a lot of it, and just about every surface is covered with the stuff. The RAT on the

sidebar 9.1 continued

two most recent Mars rovers weighs about two-thirds of a kilogram and uses only 30 watts, less power than most light bulbs. (You've got to pack light when you're traveling that far!) Yet in just two hours it can shave off a disk a few centimeters across and a few millimeters deep from even the hardest rock surfaces. Once the fresh surface is exposed, the imagers and spectrometers take over, peering through the newly formed window to analyze the rock's interior. Ah, nothing like freshly shaved rock surfaces!

existence and that its fires coincidentally died immediately before we humans started to poke around?). And if Mars is active, could there be liquid water and, perhaps, even life under its crust?

Putative (albeit still somewhat disputed) evidence for this came from NASA's orbiting *Mars Global Surveyor.* Surveying selected swaths of Mars at about 1-meter resolution, the spacecraft found what seem to be rather recent (some much less than a million years old) erosional gullies emerging from cliffs throughout the higher latitudes (fig. 9.5). On Earth such gullies commonly result from the spring runoff of snow that has accumulated on a cliff face. Because liquid water cannot exist on the surface of Mars for very long (seconds to minutes, depending on the volume and surface area), a different mechanism is likely occurring there. Perhaps the most popular of several current theories as to the origins of these gullies is that they are formed when ice plugs rupture on springs, briefly liberating a torrent of groundwater (kept liquid by geothermal heat) before the plug freezes and the flow is stopped.

This potential juxtaposition of groundwater and geothermal energy suggests that, although the surface of Mars is cold and dry today, the planet may still contain viable habitats deep within its crust. Based on this, Mars is still thought to represent the best chance for life in the Solar System outside that found on Earth. But is the Red Planet inhabited? This question motivates much current and future exploration of the planet, as we discuss in the next chapter.

The Moons of the Outer Solar System

Out beyond Mars we find the gas giants Jupiter, Saturn, Neptune, and Uranus. The physics of these planets effectively rules out any processes that we would describe as life. Jupiter and Saturn lack solid surfaces (because of the massive bulk of these planets, even their metal-and-rock cores are thought to be liquid; they haven't cooled enough since accre-

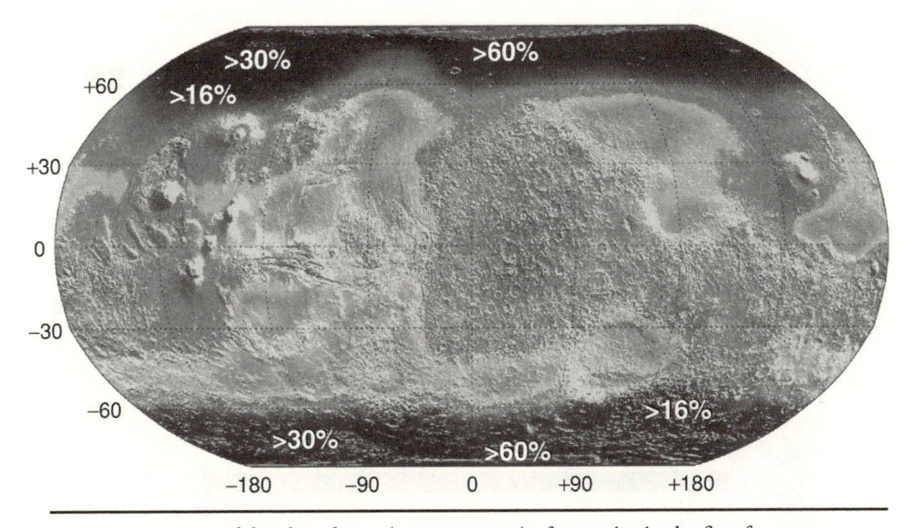

FIG. 9.4. A map of the abundance (as percentage) of water ice in the first few tens of centimeters below the surface of Mars. (Courtesy NASA/JPL)

tion to solidify), and the solid surfaces of Uranus and Neptune, if there are such surfaces, are deep below hot, highly convective seas. Life in the atmospheres of the gas giants is probably not possible (for reasons outlined in chapter 1), and life beneath the hot, turbulent seas of Uranus and Neptune is probably equally precluded by the instability of the environment. There are other environments out among the gas giants, though, that might be more conducive to life. Each of the four planets has a retinue of dozens of moons, and some of the larger of these may be among the more promising prospects for extraterrestrial life in the Solar System.

The four largest companions of Jupiter were discovered by the Italian astronomer Galileo Galilei (1564–1642) in 1610, who used them to disprove the Ptolemaic world view that all heavenly bodies revolve around the Earth. Counting from Jupiter outward, these four "Galilean satellites" are Io, Europa, Ganymede, and Callisto (table 9.1 lists some of their physical properties). While *Pioneer 10* and *Pioneer 11* flew by Jupiter in 1973 and 1974, respectively, their instruments were too primitive to return any significant data on the satellites. It was not until the two *Voyager* craft passed Jupiter in 1979 that we began to see these moons as worlds unto themselves, and much of what we now know is based on the measurements of the *Galileo* mission, which explored Jupiter and the Galilean satellites for thirty-four orbits starting in 1996. At the end of its mission in September 2003, operators at the Jet Propul-

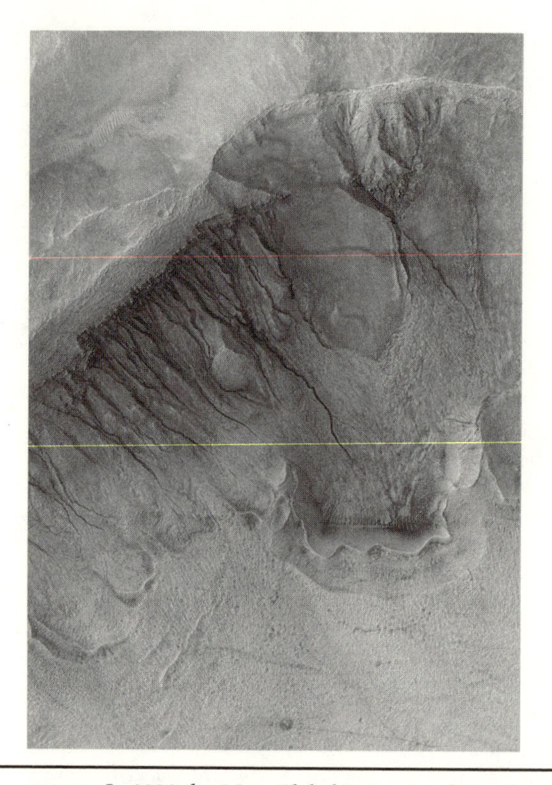

FIG. 9.5. In 2000 the *Mars Global Surveyor* orbiter photographed features, such as these, that have widely been interpreted as water-carved gullies. Whatever they are, the features seem to be very young; they lack any observable meteor craters, suggesting that they are less than a million years old. Whether this implies that Mars currently harbors liquid groundwater, however, remains contested. (Courtesy NASA/JPL/MSSS)

sion Laboratory deliberately steered the craft to crash into Jupiter in order to avoid accidentally contaminating any of the Galilean satellites with Earthly life.

One of the most exciting discoveries of the *Voyager* missions at Jupiter was stunning confirmation of a theory regarding the innermost Galilean satellite, Io. Stan Peale of the University of California, Santa Barbara, and Pat Cassen and Roy Reynolds of NASA's Ames Research Center had been pondering the ramifications of an observation made several centuries earlier: that the orbits of the inner three Galilean satellites are resonant. For every orbit of Ganymede, Europa makes two orbits and Io makes four (table 9.1). This means that the gravitational tug of Europa on Io builds up; it happens at exactly the same place in Io's

TABLE 9.1
The Galilean satellites

Name	Orbital period (Earth-days)	Radius (km)	Mass (our Moon = 1.00)	Density (g/cm^3)	Composition
Io	1.77	1,818	1.22	3.53	Rock
Europa	3.55	1,561	0.65	3.02	
Ganymede	7.15	2,634	2.02	1.94	Rock + ice
Callisto	16.69	2,408	1.46	1.83	

every orbit, forcing Io into a fairly eccentric (out-of-round) orbital path. An eccentric orbit this near Jupiter would not usually be a stable state of affairs; orbital eccentricity should lead to enormous tides in the solid material of the moon, and the friction caused as the tides flex the moon's solid rock should dissipate orbital energy until the eccentricity was damped and Io fell into a circular orbit. But the resonance with Europa prevents this; while the tides of Io try to damp the eccentricity, counter-acting tides raised in Jupiter tend to push the orbit of Io outward, causing the resonant interaction with Europa to kick up the eccentricity once again. In turn, the resonance between Io and Europa and a similar set of resonant tugs from Ganymede keep Europa's orbit out of round despite the damping tides this eccentricity causes (albeit these tides are much smaller than those observed for Io, because the amplitude of a tide drops off with the third power of the distance from the source). The net effect of all this is to convert Jupiter's rotational energy into massive tidal flexing of the crust of Io. When he worked out the precise numbers, Peale realized that so much energy is being dumped into the crust of Io that the moon should be violently volcanic—far more so than even the Earth, until then thought to be the most geologically active body in the Solar System. This purely theoretical prediction was published in the journal *Science* on March 1, 1979.

On March 8, 1979, exactly one week after the volcano prediction appeared in print, *Voyager* navigation team member Linda Morabito was looking at some images of the limb (edge) of Io. These images were over-exposed to bring out faint stars in the background for navigation purposes. Oddly, the image showed a crescent beyond the edge of the moon. At first she thought it was another of Jupiter's satellites peaking over the edge, but a quick check of the locations of the other Galilean satellites nixed this idea. Instead, the crescent was a stunning confirmation of the tidal heating theory; it was the enormous, 260-kilometer-tall plume of an extraordinarily active volcano. In short order, seven violently active volcanoes were identified on Io (including a second obvious one in the picture Morabito was using), making this moon far and away the most volcanically active body in the Solar System.

While it's a fascinating place in terms of geology, our interest in Io from the astrobiological perspective is limited; the incessant volcanism has baked Io completely dry. The same, however, cannot be said for the next moon out: Europa. Measurements of the slight deflection induced in the *Voyager* spacecraft as they flew by Europa allowed scientists to estimate the mass of this moon, which, when combined with knowledge of its size, indicated the satellite has a density of 3.02 g/cm^3. This is a little low compared with the 3.3–4.3 g/cm^3 of pure silicates, much less that of silicates with a metallic core (since iron weighs in at 7.9 g/cm^3). More perplexing still, the *Voyagers* found that Europa's is probably the smoothest surface in the Solar System; the moon is almost free of craters and its topography doesn't vary by more than a few hundred meters across its entire globe. Taken with spectroscopic evidence of frozen water on Europa's surface, these observations led astronomers to the conclusion that Europa's crust is a thick, relatively recently reworked layer of ice.

The *Galileo* orbiter added significantly to the earlier, *Voyager*-based models of Europa. During a flyby in December 1996, *Galileo*'s magnetometer measured the Europa magnetic field and found that it perfectly opposed the strong magnetic field of Jupiter. While the opposing field could have been coincidental (*Galileo* could have just happened by at the very moment when Europa's rotation brought its field into alignment with Jupiter's), follow-on flybys showed that the Europan field is always opposed to Jupiter's no matter where Europa is in its orbit. An opposing magnetic field can be generated by eddy currents, the currents induced in a conductor when it is moved through an external magnetic field. But what could the Europan conductor be? Most likely it is a global body of saltwater, a sub-ice ocean kept liquid by tidal heating; while Europa receives about one-tenth as much tidal energy as Io, that seems to be enough to keep its oceans from freezing.

While its existence is generally considered confirmed, details of the "Europan ocean" remain sketchy. *Galileo*'s gravity measurements suggest that the combined ice and water shell can be no more than 70–170 kilometers thick, but cannot distinguish the liquid from the solid phase and thus cannot estimate the thickness of the icy cover. Other, more indirect studies suggest the ice may be, at least sporadically, rather thin. For example, in February 1997 *Galileo* photographed what seem to be "ice rafts" frozen into place, as if the surface ice had once temporarily

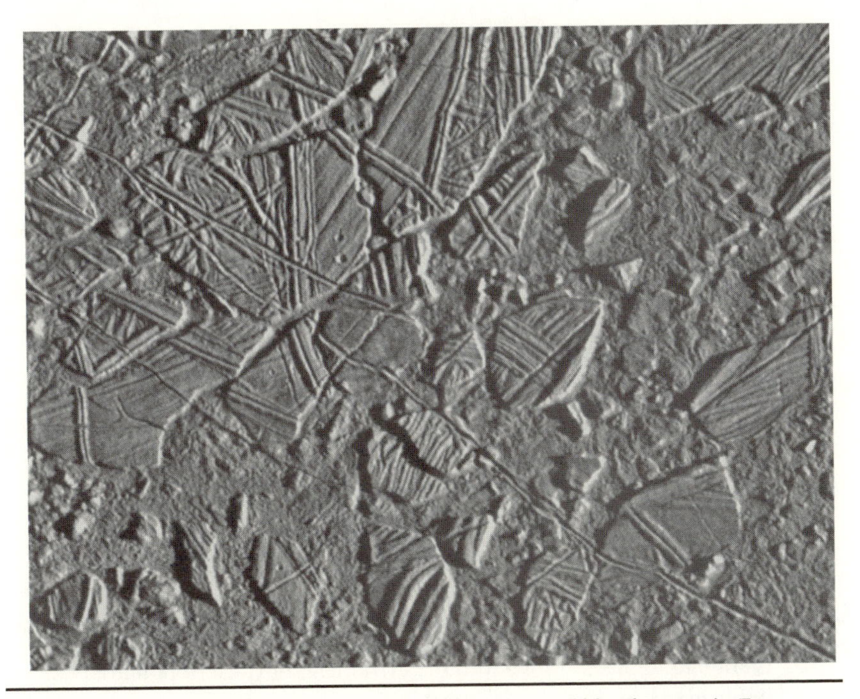

FIG. 9.6. The *Galileo* spacecraft captured this image of raftlike elements in Europa's surface ice. This has been taken as evidence that the ice forms only a relatively thin crust over the sub-ice ocean, and this crust can sometimes melt like pack ice in the Antarctic summer. (Courtesy NASA/JPL)

melted and broken into icebergs (fig. 9.6). Studies of Europa's few craters and simulations of impact crater formation in ice suggest, however, that the ice shield must be at least 3–4 kilometers thick, and may be as much as 25 kilometers thick. Thus Europa may offer a similar situation to Antarctica, where, as we described in the previous chapter, huge lakes lie hidden under many kilometers of ice.

The likely presence of a liquid ocean beneath the thick, icy crust of Europa suggests that the moon may be a potential habitat, but is it truly habitable? As we have discussed, water alone is not a sufficient criterion for habitability; we also need a source of energy to drive metabolism. Several possible sources have been suggested for Europan life. For example, even though the intensity of the sunlight that strikes Europa is only about one-thirtieth of what we receive on the Earth's surface, melt water near Europa's surface could support photosynthetic organisms. Perhaps more appealingly, the geothermal (tidal) energy that keeps Europa's ocean liquid could provide a source of energy for benthic organisms living on the ocean floor. Whether this is a sufficient source of energy, however, has been rather hotly debated in the astrobiology community. More recently a novel, Europa-specific energy source has been suggested by Chris Chyba of the SETI Institute in Mountain View, California. Chyba noted that Europa orbits deep within Jupiter's intense radiation fields and that this radiation (which would kill an unprotected human on the surface of Europa in minutes) radiolyzes ices in the moon's surface. That is, when the high-energy protons and electrons that constitute the radiation impact the ice, they tear its molecules apart and create highly reactive species. This photolysis has, in fact, been observed: spectroscopic studies of Europa from the Hubble Space Telescope indicate that this radiolysis provides Europa with a tenuous oxygen atmosphere (albeit only one-hundred-billionth as dense as Earth's). Important products of the radiolysis of Europa-like ices are the simple organic compound formaldehyde and the oxidizing agent hydrogen peroxide. From estimates of the rates with which these species are formed, Chyba has calculated that this energy source could support up to 500 tons of Europan microorganisms. This is not a huge biosphere by Terrestrial standards—the Earth's is 10 billion times larger than this—but is enough to push Europa into the extremely short list of potential habitable places in the Solar System.*

Beyond Europa we have the Galilean satellites Ganymede and Callisto. *Galileo*'s images of Ganymede have puzzled researchers, as they show two very different kinds of surface. Some 40% of the satellite's surface is so densely covered in craters that it must be billions of years old. The rest of the terrain, however, bears scars not of meteorite impacts but of extremely active tectonic movements. Both parts of the surface are believed to consist of a layer of water ice some 800 kilometers thick, but what makes the two parts so different remains to be discovered. There is no indication of liquid water, and only an extremely thin atmosphere containing the tiniest whiffs of oxygen, which again is thought to result from the bombardment of the icy surface of the moon by intense radiation from Jupiter's magnetosphere. In contrast, some magnetometer data indicate that Callisto, the furthest out of the four Galilean satellites, may also harbor an under-ice ocean. That said, an energy source sufficient to keep it liquid has not been firmly established, and the case in favor of it is generally less clear-cut than that for Europa's ocean.

*Corey Jamieson of the University of Hawaii claims we've already seen evidence for life on Europa. The closest laboratory match to the infrared spectra of the moon's surface, he says, is provided by Terrestrial extremophilic bacteria frozen to liquid-nitrogen temperatures! Still, most members of the astrobiology community are betting that a more mundane source for the Europan spectral features will be found.

TABLE 9.2
Saturn's major satellites

Name	Orbital period (Earth-days)	Radius (km)	Mass (our Moon = 1.00)	Density (g/cm³)	Composition
Mimas	0.9	200	0.0005	1.17	Rock + ice
Enceladus	1.4	250	0.001	1.24	
Tethys	1.9	530	0.009	1.21	
Dione	2.7	560	0.015	1.43	
Rhea	4.5	764	0.031	1.33	
Titan	15.9	2,575	1.83	1.21	
Hyperion	21.3	~140	0.0002	~1.2	
Iapetus	79.3	718	0.022	1.21	

Beyond Jupiter we have Saturn, with its retinue of rings and moons (table 9.2). Sadly, for all his success with the Jovian system, Galileo (now we're speaking of the Italian scientist, not the spacecraft!) did not have a sufficiently strong telescope to discover any of Saturn's satellites or, indeed, to recognize its rings for what they are.* In 1659, the Dutch scientist Christiaan Huygens (1629–95) described the rings in detail and discovered Saturn's largest satellite, Titan. Slightly bigger than Mercury, and third in size among the Solar System's satellites, Titan is currently considered the most interesting moon from an astrobiological perspective (although perhaps not in terms of its habitability). The reason for this interest is its dense atmosphere (in spite of Titan's relatively low gravity, its atmospheric pressure is 1.5 times Earth's), containing mostly nitrogen and a few percent reducing gases such as methane. Titan's atmosphere also boasts a range of simple organic molecules, including hydrocarbons and nitriles. Studying the effects of radiation on a Titan-like atmosphere, Carl Sagan of Cornell University in upstate New York concluded that the formation of organic molecules, and indeed of a mixture of high-molecular-weight solids he termed "tholins," is likely to occur on Titan. Tholins are generally red and may constitute the orange fog that renders the surface of Titan impossible to image from space using visible light.

Titan is 9.5 times farther away from the Sun than we are, and thus it

*Galileo also had the bad luck of making his observations right around the time that the Earth passed through Saturn's ring plane, rendering the rings edge-on and thus invisible. At first, it looked as if Saturn had two large companions on either side—the unresolved rings—and then months later they disappeared! One wonders what he made of all this.

receives only one-ninetieth the energy per surface area that Earth receives. This accounts for the average surface temperature of −179°C measured by *Voyager.* Sagan also argued, however, that impacts must have imparted enough energy to the moon to ensure that every part of it has seen liquid water at least during some part of its history. Could the combination of organic molecules and water on Titan create suitable conditions for life? As yet we cannot rule out this possibility, and thus the results of the ongoing *Cassini* mission to Saturn, which we touch on in the next chapter, will be of great interest. Moreover, even in the absence of any possibility of life, Titan's chemical composition may yield insights into what Earth may have been like before life took control.

And what of the rest of Saturn's retinue of four dozen (and counting) moons? The second largest Saturnian moon is Rhea, which is only one-third of Titan's diameter and one-sixtieth of its mass. Rhea is thus far too small to hold on to an atmosphere and, given its heavily cratered surface, appears geologically dead. The lack of significant atmosphere and any signs of geological activity seem to hold for the rest of Saturn's family as well, with the possible exception of Enceladus. At only 500 kilometers in diameter, Enceladus is rather small. Oddly, its surface is as white as fresh snow and is the most highly reflective surface in the Solar System. Given that impacts and "space weathering" (the cumulative effects of radiation damage) tend to darken anything exposed to space, the snow-white appearance of Enceladus suggests that its surface is constantly being replenished and, indeed, warm, water-vapor vents have been indirectly detected near its south pole. The energy source behind these vents, however, remains a mystery. But where energy is available, there could conceivably be life. Perhaps the *Cassini* spacecraft, still in orbit around Saturn, will unravel this mystery. Stay tuned.

Jupiter and Saturn, of course, aren't the only gas giants out there. Uranus and Neptune also have some middling to large moons. In particular, Neptune's largest moon, Triton, is only slightly smaller than the Earth's Moon and is the seventh largest satellite in the Solar System. At the cold, 38K temperatures found that far away from the Sun, Triton's gravity is sufficient to maintain a thin, nitrogen-rich atmosphere (only fifteen-millionths as dense as the Earth's). Intriguingly, when *Voyager 2* flew past Triton in 1989, it observed more than a dozen geyser-like plumes shooting 8 kilometers into the moon's thin atmosphere and drifting up to 150 kilometers downwind (fig. 9.7). Given that all of the plumes were observed in a narrow-latitude belt around 50° south—which during the time of the *Voyager* visit was the subsolar latitude on Triton (i.e., the region of Triton pointing to the Sun, some 4 billion km away)—it is thought that the energy source behind these geysers is prob-

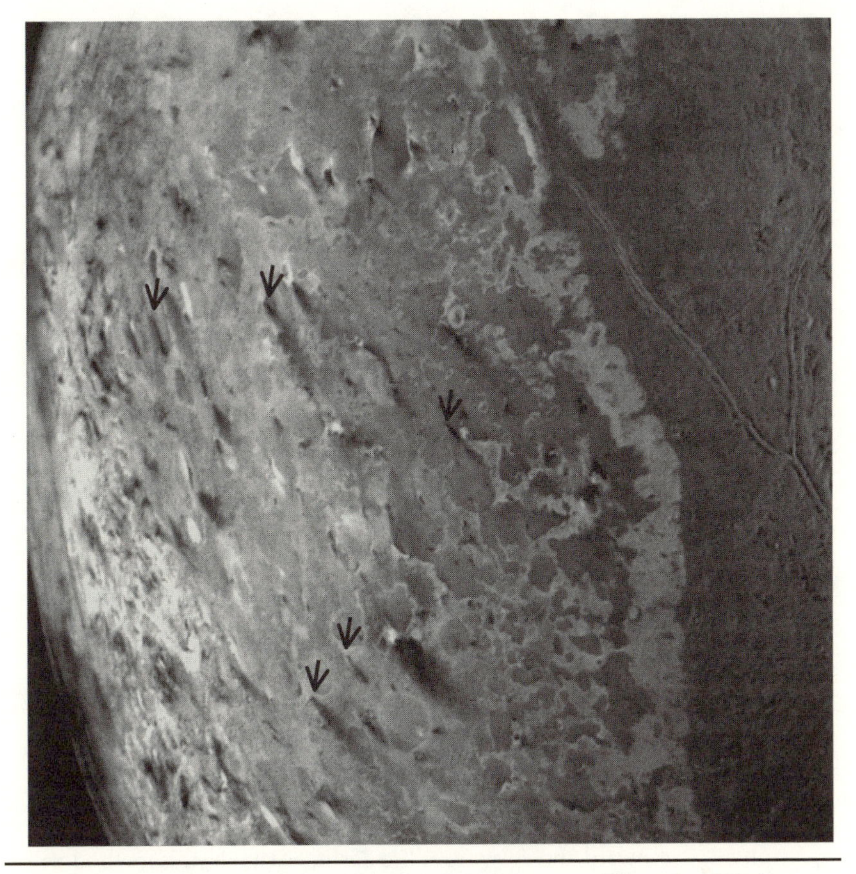

FIG. 9.7. When *Voyager 2* flew past Neptune's largest moon, Triton, in 1989, it photographed geyser-like plumes (arrows), erupting and drifting downwind. All of the plumes were observed in a narrow belt around the region of Triton pointing directly toward the Sun, so the energy source behind these geysers is probably nitrogen gas produced by solar-induced melting of nitrogen ice in the moon's crust. (Courtesy NASA/JPL)

ably solar heating. Could similar energy sources support a biosphere? Could complex, self-replicating chemistry occur at 38K? Hard to say. But, to quote J. B. S. Haldane again, "The Universe is not only queerer than we suppose, but queerer than we *can* suppose."

The Search for Planets around Other Stars

We can frequently see planets orbiting other stars on our television screens, complete with warp-driven spaceships sent from planet Earth to investigate their "strange civilizations." Sadly these are all fictional, as our civilization has so far failed to come up with a technology that would propel explorers to another star within their lifetime. The very existence of planets outside our Solar System was unconfirmed until a mere decade ago.

Until 1995, the only "evidence" for such extrasolar planets was either conjecture or very indirect. The conjecture usually revolved around arguments we've seen before: given the many billions of stars that are much like our own, it seems highly unlikely that the Sun is the only one that has planets (although the anthropic principle again weakens this argument somewhat). The indirect arguments were based on the observation that young stars are often observed to be surrounded by gas disks resembling the one from which our planetary system is thought to have formed. Thus, based on the assumption that the gas disks around young stars would tend to consolidate into planetary systems, most scientists believed there were extrasolar planets, even if they had no direct evidence of them.

All this changed in October 1995, when Michel Mayor and Didier Queloz at the Geneva Observatory announced that they had discovered a planet half as big as Jupiter flying around the yellow dwarf 51 Pegasi, in an extremely close orbit that takes just 4.2 days to complete. Eight times closer to its star than Mercury is to the Sun, this planet must endure surface temperatures of more than 1,300°C, which likely rules out life in any form or shape. Mayor and Queloz had discovered the planet by analyzing the spectra of 51 Pegasi for Doppler shifts induced when a massive planet travels around a star at neck-breaking speed, alternatively pitching it toward and away from the Earth. Naturally, this method will detect only large planets at very short distances from their stars. The effect that a small, relatively distant planet like ours has on the motion of its star is orders of magnitude smaller and cannot be measured using current astronomical methods.

The first discovery of an extrasolar planet unleashed a kind of gold rush. Within weeks, other astronomers had confirmed the evidence for the planet around 51 Pegasi and came up with two further candidate stars with suspicious Doppler shifts. Within a few years, the list of extrasolar planets grew to more than a hundred. By 2003 somewhat more direct evidence of an extrasolar planet was achieved when one of them was seen to dim the light of its host star as it transited (passed in front of the star as observed from Earth). While observing transits is, in theory, a more sensitive method for detecting extrasolar planets than the Doppler method, transits are so rare that large amounts of observation time and automated data analysis are required to detect planets via this approach.

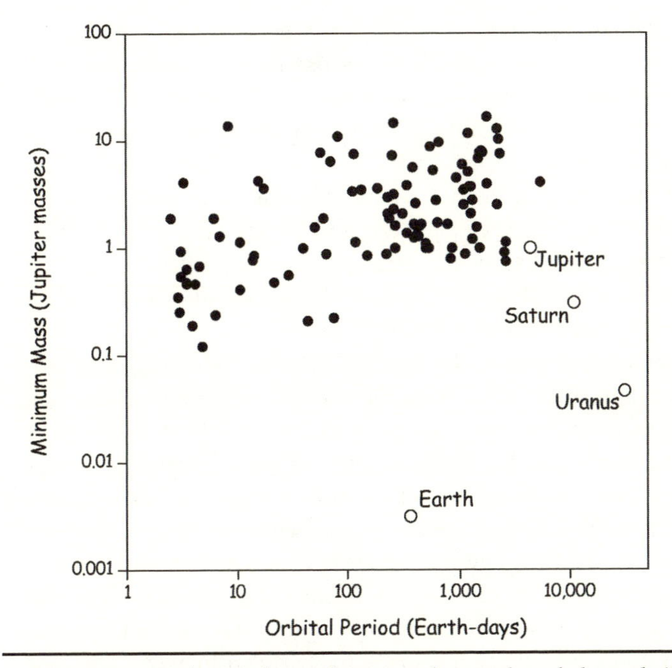

FIG. 9.8. Because larger, closer planets tend to push and shove their host star more than do smaller, more distant planets, all of the first hundred extrasolar planets discovered are enormous and in close, rapid orbits, as shown here. The mass and orbital period of various planets in the Solar System are shown for comparison. No Earth-like planets have yet been identified.

So far the "extrasolar planet encyclopedia" is filled with gas giants in ridiculously small orbits (fig. 9.8). For example, the largest known extrasolar planets are ten times the mass of Jupiter, and even the smallest known extrasolar planet is still ten times the mass of Earth. This does not imply, however, that most planets are of Jovian proportions and in close, Mercury-like orbits; it simply reflects two limitations in the current data. The first is that larger planets push and shove their companion stars harder than smaller planets do, and thus the larger the planet, the easier it is to detect via Doppler shifts in the star's spectrum. The second is that, to be compellingly identified, a planet's effect on its star must be observed for at least one full orbit. Given that the longest searches started just a decade ago, it is not surprising that all of the extrasolar planets discovered to date orbit their companion stars in less than 8.5 Earth-years. With increasingly long surveys, more gas giants in distant Jupiter- and Saturn-like orbits will no doubt be observed. Observation of *smaller* planets, however, is another matter, although several instruments for the detection of Earth-sized planets have been proposed or are even under development—which we discuss in the next chapter.

And might any of these extrasolar planets harbor life? It is, one must suppose, a matter of probabilities. One must multiply the probability of life arising on a suitable planet by the number of suitable planets. And given that the number of identified extrasolar planets is rapidly increasing, there must be many such planets, right? Perhaps. But a critical evaluation of the criteria that seem to be required suggests that the number of planets that might be able to support life may be much smaller than the optimistic, early predictions.

An issue we have already discussed in detail (chapter 3) is the problem of continuously habitable zones. Given the uniquely life-supporting properties of water, we should probably assume that life forms only in zones around stars in which water can remain liquid. Similarly, given our assumption that life takes time to arise and evolve, we'd best limit our speculations to planets that remain in this liquid-water, "continuously habitable zone" over geological timescales. This is not a trivial issue; in the Solar System, the continuously habitable zone contains only one planet, and that planet has only just managed to stay within the zone's limits over its 4.5-billion-year history. If the Earth's orbit were even a few percent smaller or larger, the oceans would have boiled under the influence of the slowly increasing brightness of our Sun (which, again, is 20% brighter than it was 4 billion years ago) or would have remained frozen for most of the history of the Solar System. And as we discussed at length in chapter 2 (see sidebar 2.3), only about 1 in 1,300 stars seems as well suited as our Sun to host inhabited planets. When we couple this strong constraint with our poor knowledge of how frequently rocky planets form (since we can't yet detect small, Earth-like planets), we are forced to admit that we simply do not know how frequently terrestrial planets form within the habitable zones of suitable stars.

Conclusions

So what is the bottom line on habitable places in our Universe? It is clear that, in the Solar System, neither Venus nor Mars resides in the continuously habitable zone. And while Mars (and perhaps even Venus) may have hosted liquid water once upon a time, the surfaces of both planets are far too dry (and cold and hot, respectively) to support life now. Still, big questions remain. Did life arise on Mars when it was more clement and then, perhaps, flee to still living habitats beneath the surface? And when nonsolar energy sources (e.g., tidal heating) are available to keep things warm and moist, can life arise in places, like Europa, that fall far outside the classical habitable zone? And what of life around other stars?

In the next chapter we detail humanity's efforts to answer these exciting questions by searching for evidence of life in these far-flung places.

And what about those early *Apollo* astronauts? Obviously they—and the rest of the biosphere—survived. Six months after the *Apollo 11* mission, the astronauts of *Apollo 12* were similarly quarantined. When they emerged from their three-week isolation unscathed, the decision was made that the Moon posed no threat and the astronauts of the four remaining *Apollo* missions were not subjected to the same isolation. Not that these "planetary quarantine" issues are behind us, though. While the planned date of NASA's Mars sample return keeps moving farther away at more than one year per year (i.e., it is receding rapidly into the distant future), someday, we presume and hope, we will face the decision about quarantining samples from our neighboring planet. As far back as 1997 a National Research Council report argued that, although the probability that such samples would contain pathological or environmentally dangerous organisms is low, we should not simply assume the risk is zero. The consensus then and now is that such samples should be delivered to a combined quarantine and research facility unlike any other in existence—one that is capable of protecting the scientific integrity of the extraterrestrial samples (i.e., preventing their contamination with Terrestrial substances), while at the same time protecting Earth's environment from exposure to potentially dangerous organisms from Mars.

Further Reading

The exploration of Mars. Morton, Oliver. *Mapping Mars.* New York: Picador USA, 2002.

The Galilean satellites. Showman, A. P., and Malhotra, R. "The Galilean satellites." *Science* 286 (2003): 77.

Europa as a potential abode for life. Pappalardo, R. T., Head, J. W., and Greenley, R. "The hidden ocean of Europa." *Scientific American,* October 1999, 54–63.

Extrasolar planets. Encyclopedia of extrasolar planets, www.cfa.harvard.edu/planets.

CHAPTER 10

The Search for Extraterrestrial Life

The spacecraft was the most sophisticated robot that its creators could muster. Traveling through the planetary system of a yellow dwarf star a third of the way out from the center of the Milky Way, it passed just 1,000 kilometers away from the surface of a small planet (some 12,000 km in diameter). Visually, the planet appeared somewhat unusual. For example, at the resolution of the onboard cameras (tens of meters to kilometers) fully three-quarters of the surface looked perfectly flat. The remaining one-quarter of the planet was quite rugged but, unlike nearly every other solid body in this planetary system, devoid of any obvious impact craters. Clearly erosion was filling in craters much faster than the rate at which they were forming. Much of the rougher quarter of the planet's surface, especially those portions near the equator, were covered with a large quantity of something that was strongly absorbing red light. The onboard infrared spectrometer detected water vapor in the planet's atmosphere at concentrations that would be saturating, given the planet's temperature (which the probe's spectrometers had determined to be ~20°C); maintaining such saturating conditions would require significant reservoirs of liquid water in equilibrium with the atmosphere. Could the flat, blue-green areas on the planet be this reservoir? The same onboard spectrometers detected some serious chemical abnormalities in the atmosphere, such as a large amount of molecular oxygen. They also found the clear spectral fingerprints of a fraction of a percent of carbon dioxide, and small but significant, parts-per-million quantities of methane as well.

The question on everyone's mind was, not surprisingly, did the planet harbor life? The high oxygen content of its atmosphere hinted that it might. Oxygen is extremely reactive; it is the second most strongly "oxidizing" element in the periodic table, and thus is unlikely to remain in a planetary atmosphere unless it is constantly replenished. This is all the more true when, as was the case here, geology is constantly churning the surface of the planet and exposing fresh rocks for the oxygen to react with. Still, oxygen alone is not proof of life; abiological processes also produce this reactive gas, though they would be stretched to produce so much of it. Much more telling was the simultaneous presence of oxygen and methane. Given the avidity with which the two react, even the trace amounts of methane detected were a hundred orders of magnitude higher than would be expected at chemical equilibrium (i.e., at equilibrium, not a single methane molecule should exist in the planet's atmosphere). Moreover, the half-life of methane in an oxygen-rich atmosphere at these temperatures is at most a decade or two, suggesting that something was rather rapidly replenishing the planet's methane supply. This massive, actively maintained disequilibrium seemed the surest signature of life on this rocky planet—perhaps associated with the red-absorbing pigment?

And if there was life, was there intelligent life? The imagers aboard the spacecraft saw no signs of roads or cities or other artificial creations. (Passing over the daylight side of the planet, it did not have an opportunity to look for artificial lights.) But its radio-science instruments, built to monitor the motions of charged particles in planetary magnetic fields, recorded some powerful and unusually regular radio pulses—a sign of intelligence? The case for this was less clear than the case for life itself, but even the skeptics had to wonder.

Looking for Life

Considering our natural environment with its meadows, trees, and large animals, a naive observer might be forgiven for thinking that life should be easy to detect. But these highly visible, macroscopic forms of life appeared only relatively late in the evolution of life on Earth. During the first 70% or 80% of its multibillion-year residency on Earth, life was represented exclusively by microbes. And in many extreme environments, microbes still are the only forms of life. In many cases, finding and identifying these bacteria and archaea remains a serious technical challenge to microbiology. For example, analysis of large collections of DNA randomly collected from samples of ocean water indicate that more than 90% of the bacteria in any given sample are species that have never been cultivated in the laboratory and formally identified. It is probably a good bet that an even smaller fraction of the myriads of microbes living in cracks in rocks deep beneath the Earth's surface, or perhaps even suspended in clouds, have been identified to date. And if it is so hard to identify these organisms here on our own planet (chapter 8), how feasible will it be to identify life elsewhere?

There are both obvious and not-so-obvious reasons that the identification of extraterrestrial life will be difficult. At the obvious end of the spectrum, we have the problem that opportunities to bring extraterrestrial samples into the laboratory, where they can be subjected to intensive study, are rather rare (although they do exist, as we'll show later). A second reason is perhaps less obvious, but much more profound: how would we recognize extraterrestrial life if we saw it? Given the far-reaching biochemical homologies between all Terrestrial species, it is clear that all life on Earth shares a common origin. Because all known Terrestrial life employs, for example, ribosomes, the search for life in new environments on Earth (once the possibility of contamination has been excluded) can thus be reduced to the detection of novel ribosomal RNA genes. The assumption that extraterrestrial life shares this trait, however, is extremely weak at best; after all, the biochemistry of a truly alien life form may not be at all similar to ours. So, then, how do we detect life while making the fewest possible assumptions about what the chemistry of that life will be? This is not an easy question to answer. To delve into it, let's look at how it has been done (on the few occasions when it has been done) in the past, and what has been proposed for the future, starting with our neighboring planet that has so often hosted imaginary life forms.

The Search for Life on Mars

While orbiters, such as *Mariner 9* and the more recent *Mars Global Surveyor,* have been invaluable in the charting of Mars topography, the detailed search for past or present life focused on landing missions. The Soviets were the first to attempt to land craft on the Martian surface, but failed to get scientific data back from any of them. In November 1971, for example, the landing probe of the *Mars 2* mission suffered a catastrophic failure and crashed, becoming the first human-made object to reach the surface of the Red Planet, but returning no data. A few days later its sister craft, *Mars 3,* touched down softly, but fell silent 20 seconds later; equipped with a landing program that could not be altered during its approach to the planet, it landed in the middle of a major sand storm and lost contact. Two years later the *Mars 6* lander stopped transmitting some 12 kilometers above the Martian surface, and *Mars 7* accidentally released its lander 4 hours early, putting it in a solar orbit that missed Mars by 1,300 kilometers. The first close-up look at the surface of Mars thus didn't occur until 1976, when the twin American spacecraft *Viking I* and *Viking II* made soft landings there. The first images after touchdown showed nothing more than sweeps of rocky red desert, but as the first images ever taken from the surface of another planet, they were greeted with enthusiasm and graced the front pages of newspapers around the world.

Exciting as the first images from the surface of Mars were, perhaps still more exciting were the *biology* experiments the twin landers were to

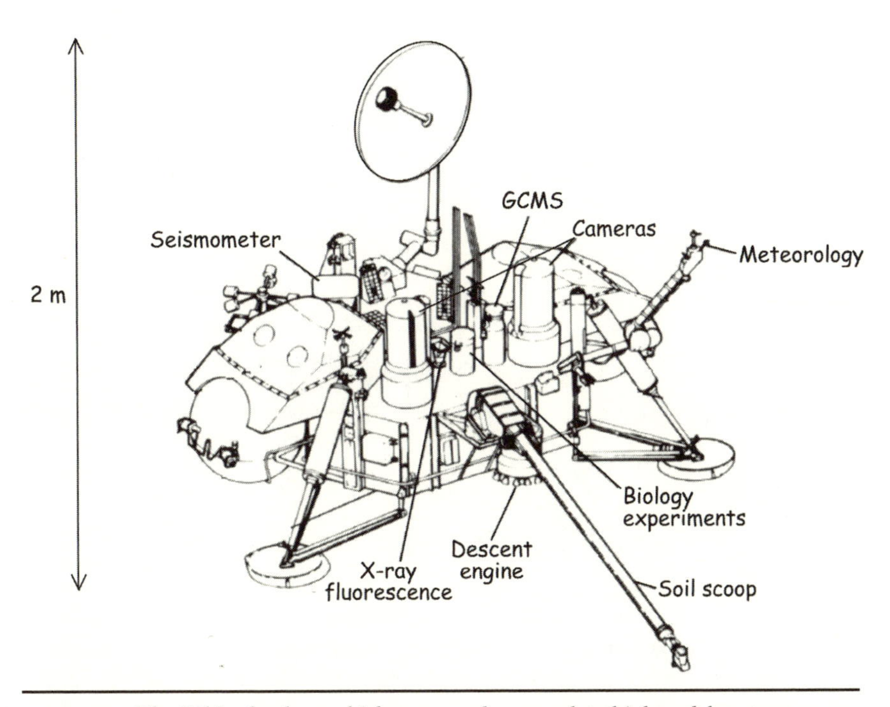

FIG. 10.1. The *Viking* landers, which crammed a complete biology laboratory into a few liters of space and a few kilograms of mass, represent the most sophisticated search-for-life study conducted by humans to date. (Courtesy NASA)

conduct starting a few days after arrival. The life-search systems of the *Viking* missions, developed by a large, highly interdisciplinary team led by NASA, included three different biochemical experiments (really an entire, state-of-the art analytical laboratory crammed into a few liters of space on a mass- and power-limited spacecraft; see fig. 10.1), aimed at using radioactive markers to detect carbon-based metabolism. Each experiment was carried out with multiple "fresh" and heat-"sterilized" samples, the latter acting as a control experiment.

The first *Viking* biology experiment, called the labeled release (LR) experiment, was headed by Gilbert Levin, a one-time "sanitary engineer" in California who developed techniques to detect bacterial contaminants in drinking water and then went on to become a prominent astrobiologist. The LR experiment was aimed at detecting catabolic metabolism (fig. 10.2). That is, metabolism in which organic molecules supplied as food are broken down into still simpler, lower-energy carbon compounds that are then "exhaled" into the environment. In the LR

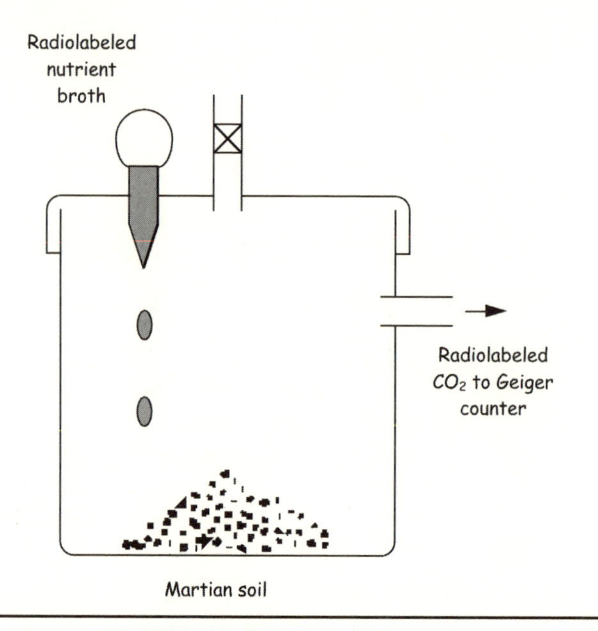

FIG. 10.2. The *Viking* labeled release (LR) experiment searched for catabolic metabolism—that is, the breakdown of simple, fixed-carbon compounds supplied in a nutrient broth to carbon dioxide or other gases. The experiment makes the perhaps fundamental assumption that Martian bugs eat the same sort of simple organic molecules that Terrestrial bacteria consume.

experiment, small amounts of an aqueous "nutrient broth" were added to a sample of Martian soil. The nutrient broth contained small organic molecules, including formate, glycolate, glycine, alanine, and lactate, which had been labeled with the radioisotope carbon-14 (^{14}C). The thought behind the experiment was that Martian bugs in the soil would eat the labeled nutrients and exhale radioactive carbon dioxide, carbon monoxide, or, perhaps, methane. The atmosphere over the soil sample was tested using a Geiger counter to detect any radioactive gases released.

The initial LR results radioed back to Earth seemed to support the idea that the Martian soil contained life (fig. 10.3). Specifically, after the first addition of nutrient broth to a "fresh" soil sample, the level of radioactive gas slowly rose before leveling off after several days. Heat-treating a sample at 160°C abolished the effect and no labeled gas was observed. Heating to 50°C did not entirely abolish the effect, but led to very significant decreases in gas output. Similar effects had been seen with Terrestrial soils, with gas production leveling off after a few days when the bacteria in the sample had consumed all of the added nutrients. Terres-

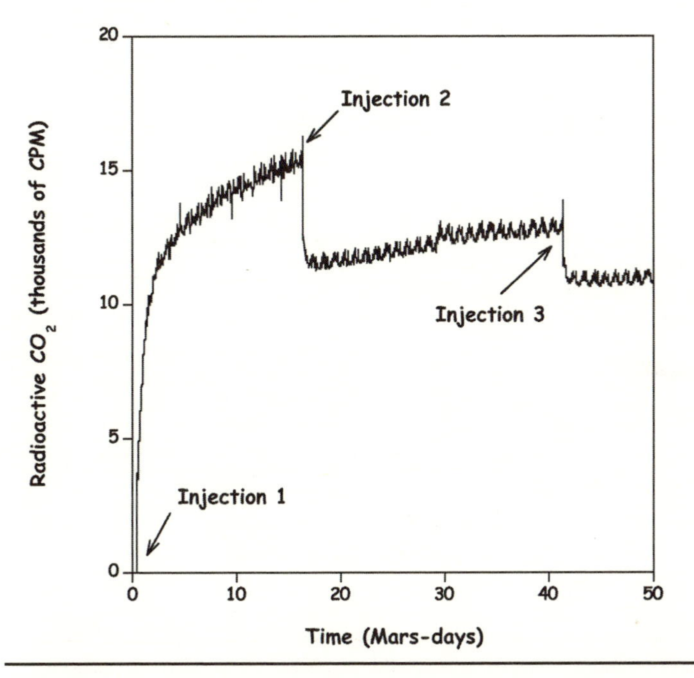

FIG. 10.3. The *Viking* LR results seemed at first to suggest that the Martian soil contains life: after the initial addition of radiolabeled nutrient broth, the level of radioactive gases in the chamber (measured in counts per minute, CPM) rose slowly for several Mars-days before leveling off. With Terrestrial samples, however, a second injection of nutrients had led to the production of more radioactive gas as the dormant bacteria began to consume this new dose of food. This was not true of the Martian soil: the second and third nutrient injections did not produce any further release of labeled gases.

trial samples similarly failed to emit gas if sterilized at 160°C and produced much less gas if first heated to 50°C. With unsterilized Terrestrial samples, though, the addition of more nutrients after the initial incubation would then produce still more radioactive gas as the dormant bacteria sprang into action to consume the new dose of food. This was not true of the Martian soil; on Mars, the second and third nutrient injections did not produce any further release of labeled gas.

The pyrolytic release (PR) experiment, headed by Norman Horowitz (1915–2005) of the California Institute of Technology, also aimed at carbon metabolism, but this time researchers were looking for metabolism running in the opposite direction: anabolic metabolism—organisms taking up carbon dioxide (or carbon monoxide, a small amount of which is present in the Martian atmosphere) and producing higher-

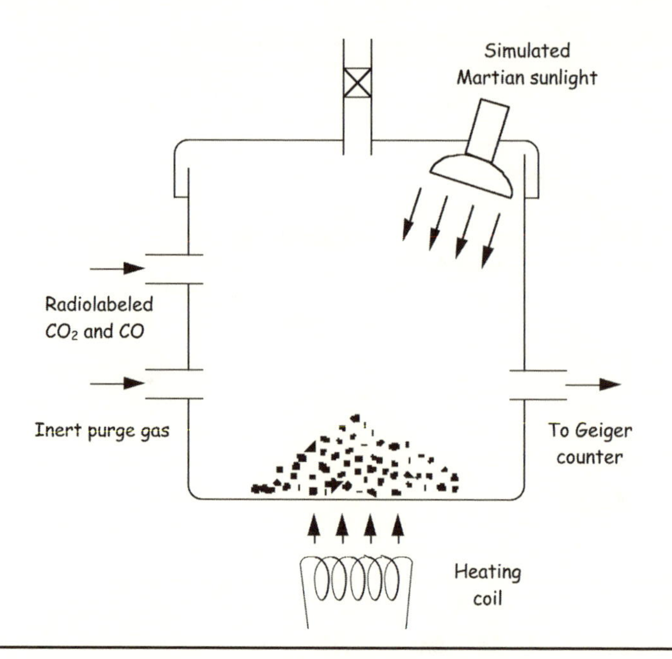

FIG. 10.4. The *Viking* pyrolytic release (PR) experiment searched for anabolic metabolism—that is, the fixing of carbon dioxide or carbon monoxide into reduced carbon compounds. After incubation with ^{14}C-labeled CO_2 and CO (both in the dark and, in the hope of encouraging photosynthesis, in simulated sunlight), the soil was heated to 635°C to pyrolize (char) any fixed carbon into volatile compounds, which could then be detected by a Geiger counter.

molecular-weight carbon compounds of their own (fig. 10.4). Martian soil samples were placed in the test chamber and incubated for 5–139 days in the presence of ^{14}C-labeled carbon dioxide and carbon monoxide in a ratio close to that observed in the Martian atmosphere. Given that the most obvious carbon-fixing metabolism on Earth is photosynthesis, the experiments were run with and without a light mimicking the Sun. After the incubation period, the chamber was purged with helium to evacuate the labeled CO and CO_2, and then heated to 635°C to pyrolize any fixed carbon compounds into volatile gases. Any radiolabeled gases released were then detected using a Geiger counter.

In total, the two *Viking* landers investigated nine soil samples by using the PR experiment. After incubation, seven of the nine experiments produced detectable peaks in the quantity of radioactive carbon atoms released from the soil on heating (fig. 10.5). The peaks in question were small (roughly equivalent to the metabolic intake of 1,000 bacterial

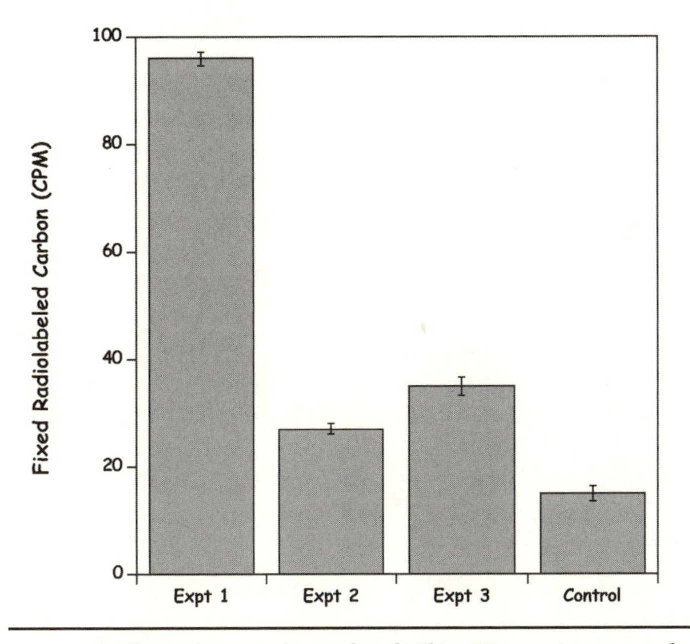

FIG. 10.5. Shown here are the results of *Viking* PR experiments conducted with four different soil samples from the *Viking 1* site. Small and varying, but statistically significant (error limits indicated by antennae at the tops of the bars), amounts of fixed carbon were detected in three experiments. A fourth, control sample was heated to 175°C for 3 hours before incubation with the radiolabeled gases. It produced a significantly smaller signal.

cells), but they were much larger than the peaks seen for sterile Terrestrial soil samples, suggesting that the experiment had indeed detected life. In contrast, Martian soil samples that were sterilized by heating to as little as 50°C before addition of the radiolabeled gas did not produce any significant peak, again consistent with a biological interpretation of the PR results. But how firm was the conclusion that the PR experiment had detected life? A weakness was that the conditions under which the PR experiment detected life were different than those of the LR experiment. The putative organisms in the LR experiment were not killed when heated to 50°C, whereas the putative organisms responsible for the PR result were killed at far lower temperatures.

The third *Viking* biology experiment was the gas exchange (GEx) experiment, headed by Vance Oyama of NASA's Ames Research Center in California. This was also aimed at the gaseous products of metabolism, but used a gas chromatograph / mass spectrometer pair (GCMS) to detect not only CO_2 (as in the LR experiment) and fixed carbon (as in the

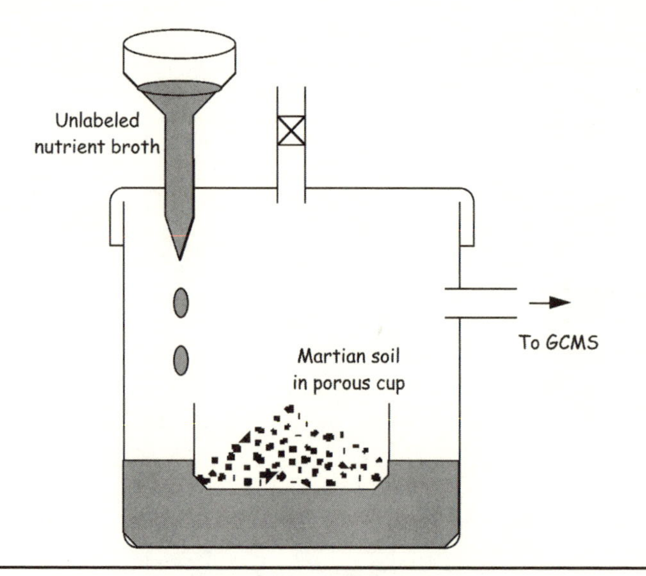

FIG. 10.6. The *Viking* gas exchange (GEx) experiment measured the gaseous products of metabolism. It used a gas chromatograph/mass spectrometer pair (GCMS) to detect not only CO_2 and fixed carbon but also metabolic products lacking carbon, such as O_2, H_2 or N_2. The soil sample was introduced into the test chamber, and the chamber's atmosphere was sampled. The soil sample was then moistened with a little water vapor ("humid mode"), incubated, and the atmosphere reexamined; any differences between the two samples were interpreted as a sign of activity. After several Mars-days, enough aqueous nutrient broth was added to the chamber to wet the soil ("wet mode," as shown here), and the gas analysis was continued.

PR experiment) but also metabolic products lacking carbon, such as O_2, H_2, or N_2 (fig. 10.6). First a soil sample was introduced into the test chamber and the chamber's atmosphere sampled. The soil sample was then moistened with a bit of water vapor ("humid mode"). After a period of incubation, the atmosphere in the chamber was reexamined, and any differences between the two atmospheric samples was viewed as a sign of activity. Later, a significant amount of aqueous nutrient broth was added to the chamber such that the soil became physically wet ("wet mode"), and the gas analysis was continued. As a control experiment, some soil samples were "sterilized" at 145°C before analysis.

The results of the GEx experiment were surprising (fig. 10.7). In the humid mode, a reasonable quantity of oxygen emerged from the soil sample, more or less immediately, but the rate of oxygen production promptly slowed down. In contrast to the humid-mode experiment,

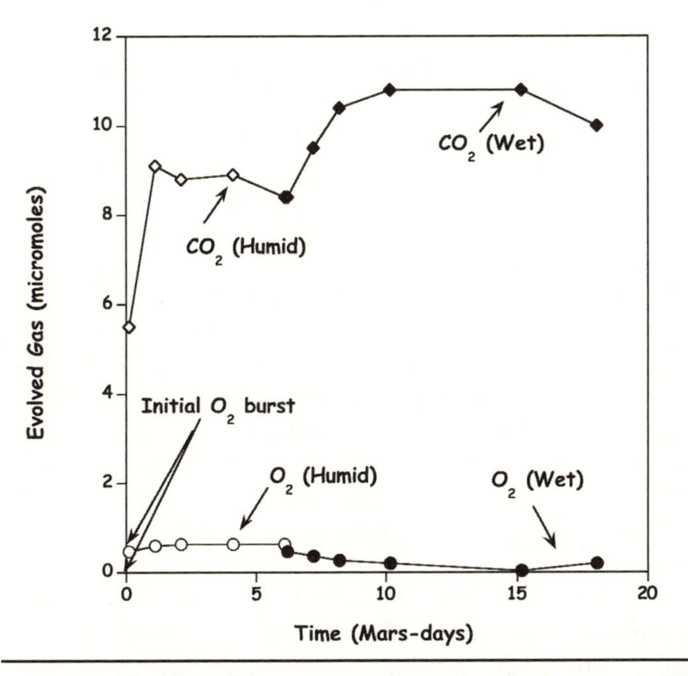

FIG. 10.7. The *Viking* GEx experiment detected modest increases in CO_2 (upper line) and a large surge of O_2 (from initially undetectable levels) as soon as water vapor was added to the soil-containing chamber. But the rate of O_2 production promptly slowed down. After six Mars-days, when the sample was wetted with "broth," no additional O_2 was produced. In fact, the O_2 levels initially present dropped. Was this consistent with life? Probably not. The initial O_2 release occurred in a sudden burst, whereas bacteria-laden Terrestrial samples first release gases slowly, then more rapidly as the bugs multiply. Perhaps more tellingly, heat-treated Martian soil samples produced just as much O_2 as untreated samples.

however, no additional oxygen was observed in the wet mode. In fact, the trace amount of oxygen initially present in the chamber dropped, suggesting that oxygen was being absorbed by the soil. Was this consistent with life? The humid-mode production of oxygen was clearly reminiscent of photosynthesis. But it happened in the dark! Moreover, the oxygen was released in a big, sudden burst; bacteria-laden Terrestrial samples, in contrast, start releasing gases slowly and then more rapidly as the bugs multiply. Perhaps most tellingly, heat-treated Martian soil produced just as much oxygen as untreated samples.

What does this jumble of conflicting results mean? For most of the *Viking* scientists, the final conclusion was that the missions failed to detect life in the Martian soil. In large part these doubts were driven by the results of yet another *Viking* experiment: a detailed investigation of the soil composition using GCMS. This experiment utterly failed to uncover any organic molecules, even at the parts-per-billion level (the instrument gain was turned up so high that the scientists finally detected traces of the solvents used to clean the chamber a year earlier when it was still in Florida). For a soil sample on Earth there are at least traces of thousands of dead organisms for every living one, and thus life is always associated with large amounts of carbon compounds. But, if "no carbon" is the equivalent of saying "no life," what accounts for the seemingly positive LR and GEx results? Could they just be the result of some odd abiological chemistry occurring on the cold, dry surface of the Red Planet?

Hints as to the possible abiological origins of the *Viking* results have come from studies of soils on Earth. Many Terrestrial soils (even sterile ones) take up carbon dioxide when warmed (and the *Viking* biology experiments were conducted at temperatures higher than the ambient temperature on Mars), and this adsorbed carbon dioxide can often be liberated by heating. Thus simple physical processes readily account for the results of the PR experiment. But what about the LR and GEx experiments? They too might be chemistry. For example, Albert Yen of the Jet Propulsion Laboratory has shown that, under extremely cold and dry conditions, ultraviolet light (remember: Mars lacks an ozone layer and thus the surface is bathed in UV) can cause carbon dioxide to react with soils to produce a highly reactive class of oxygen compounds called superoxides. When mixed with small organic molecules, superoxides oxidize them to carbon dioxide, thus possibly accounting for the LR result. Superoxide chemistry could also account for the puzzling results seen when more nutrients were added to the soil in the LR experiment; because life multiplies, the amount of gas should have increased when a second or third batch of nutrients was added, but if the effect resulted from a chemical that was consumed in the first reaction, no new gas (as observed in the experiment) would be expected. Lastly, many superoxides are relatively unstable and are destroyed at elevated temperatures, thus also accounting for the "sterilization" seen in the LR experiment. Superoxides might also explain the GEx humid-mode result: many superoxides react with water to produce oxygen. And so, while several of the *Viking* experiments produced results that, before the mission's launch, had been considered tell-tale signs of life, and while a few prominent scientists still argue the case for life on Mars, most experts now begrudgingly concur that the *Viking* results were negative.

The *Viking* biology results are thus best summed up as a cautious "no life has been found." But even if we assume the experimental results truly are negative, what does this really mean in terms of life on Mars? The *Viking* experiments, like almost everything in the search for extraterrestrial life, were prejudiced by what we know about life here on Earth. The landers, for example, were designed to look for carbon-based life (as we touched on in chapter 1, this is not an unreasonable assumption), exposed on the surface of the planet (perhaps a poorer assumption, given the intense solar UV and cold, dry conditions of the surface). And not just any carbon-based life; the labeled release experiment assumed that we know something about the Martian bugs: that they eat the same simple carbon compounds loved by Terrestrial bacteria. Likewise the gas exchange experiment assumed that photosynthesis occurs on Mars as well. While these may (or may not) be reasonable assumptions, one has to be aware that any such assumption constrains our chances of finding extraterrestrial life.

Post-*Viking* Exploration of Mars

Although the *Viking* probes continued to send back data to Earth for more than five years, and thus stand out as a major success story among the many failed Mars missions, the disappointment of their biology experiments, along with the measurements that suggested the general conditions were too hostile to support life, discouraged further Mars exploration for almost two decades. Only after the discovery of chemosynthetic food webs in deep-sea and underground ecosystems here on Earth (chapter 8) did scientists realize that life on Mars might in fact be hidden underground, or within the ice caps. By the mid-1990s, the astonishing discoveries of life in extreme conditions on Earth helped scientists to regain some optimism about the possibility of past or present life on Mars.

Inspired in part by this new perspective on the history of the Red Planet, NASA's Mars program resumed in the early 1990s, but it suffered an immediate setback in August 1993, when the probe *Mars Observer* disappeared as it attempted to brake into Mars orbit, when it is presumed to have exploded. In response to the loss of this billion-dollar, multiyear mission—and more problems: a crippling antenna problem on the otherwise successful, decade-long Jupiter project *Galileo,* and the initially near-sighted mirror launched in the multibillion-dollar Hubble Space Telescope—NASA launched a new initiative termed Discovery Missions, which were intended to be "faster, better, cheaper." Arriving at Mars in 1997 after only a few years of design, fabrication, and testing (and for a relatively paltry $154 and $256 million, respectively), the *Mars Global Surveyor* orbiter and the *Mars Pathfinder* lander opened a new era of Mars exploration, which is still going on. While neither of these missions was specifically designed to look for traces of life, they both laid important foundations for the later explorations of the planet and search for life.

Follow-on missions run by NASA as part of the quest to understand whether Mars has, or had, what it takes to harbor life have focused on the mantra "follow the water." As described in the previous chapter, *Mars Odyssey* orbiter carried a neutron spectrometer that detected abundant low-energy neutrons, signifying hydrogen, no doubt as water, on the Martian surface. And the Mars Explorations Rovers, *Spirit* and *Opportunity*, are, at the time of this writing (more than one Mars-year later), roaming the Martian surface using spectroscopy and imaging to probe for geological signs of past water. But none of these missions are equipped to detect the signatures of life itself. Instead, these missions are aimed primarily at improving our knowledge of Martian geology and climate history, thus providing a clearer picture of where the remnants of Martian life might be hiding.

In contrast, the European Space Agency's Mars probe, the *Mars Express* orbiter, has been searching for specific signatures of life. Using the most sophisticated remote sensing spectrometers sent to the planet to date, *Mars Express* has detected traces of methane in the Martian atmosphere. Notably, there are hints that the methane is not equally present everywhere on the Red Planet; it seems to be concentrated over the areas that *Mars Global Surveyor* has indicated to be rich in water (presumably as ice). But is this an indication of life? It does represent the sort of disequilibrium that might be associated with life: while the amount of methane detected is quite small, only about 10 parts per billion, even the Martian atmosphere is reactive enough that the equilibrium concentration should be far, far lower. Indeed, the half-life of methane in the Martian atmosphere should only be around 600 years (carbon dioxide and methane react to form water and carbon monoxide). Taken together, these observations mean that, to maintain even the small concentration that is observed, something must be actively pumping out about 150 tons of methane per year. But must this "something" be life? On Earth methane is primarily the product of life,* but geological processes can also produce it from abiotic precursors such as water, carbon dioxide, and hot olivine, an igneous, reducing mineral that is known to be reasonably plentiful on Mars. In fact, while it is not common, the gas exhaled by at least some Terrestrial volcanoes is up to 0.1% methane. Thus the jury's still out on whether the methane on Mars is a signature

*Of note, 150 tons per year is the amount of methane produced by about 1,000 cattle. This allows us to put a firm upper limit on the bovine population of the Red Planet.

of Mars biology. The key unanswered question here is whether Martian geology is active enough to produce the observed amount of methane. As we described in the previous chapter, images taken by the same *Mars Express* orbiter indicate some of the Martian volcanoes have been active in the last few million years, suggesting that Mars remains at least somewhat geologically active today. Thus, sadly, we cannot rule out dull old geology as a source of the observed methane.

The *Mars Express* orbiter also carried a lander with it, the *Beagle II* (named after HMS *Beagle*, the ship that Darwin sailed on), which fell silent while attempting to land on Christmas Day 2003. The lost lander was a stationary device (as opposed to the mobile, U.S.-led rovers), which would have explored the Martian soil and shallow subsoil at its landing site for traces of past or present life. Had the *Beagle* succeeded, it would have provided precise measurements of the carbon and sulfur isotopes on the Martian surface, both of which (as we discussed in chapter 7) can contain potential life signatures.

And next? Interest in Mars continues at a fairly high level; over the next decade NASA plans to send another orbiter, a nonmobile, high-latitude lander, and a much more sophisticated rover. The orbiter, *Mars Reconnaissance Orbiter*, is slated to arrive at Mars in March 2006 and will carry with it by far the highest-resolution cameras (capable of resolving surface features as small as 30 cm) and a high-resolution imaging spectrometer that should be wonderful for spotting the signatures of—to speculate a bit—hydrated minerals surrounding some ancient hydrothermal vent. In May 2008 the *Phoenix* lander (a follow-on to the failed *Mars Polar Lander*, which crashed in late 1999, and named, of course, after the legendary bird that is reborn from its own ashes) will land near the icy North Pole region of Mars between latitudes 65° and 75° north, to investigate the possible permafrost identified remotely by *Mars Global Surveyor* (see fig. 9.4). The *Mars Science Laboratory*, which is currently planned for an October 2010 landing, will be the most sophisticated rover yet placed on any extraterrestrial body. Unlike previous Mars rovers, this one-ton mobile robotic laboratory will collect and crush rock and soil samples and distribute them to sophisticated onboard instruments for detailed chemical analysis. And while the precise instrument payload is not yet defined, it is planned that the craft will carry a suite of instruments aimed at detecting and identifying organic molecules, such as amino acids and nucleobases, and identifying atmospheric gases that may be signatures of biological activity.

Have Martians Landed on Earth?

While the in situ exploration of Mars seems to be going well, plans for a mission that collects samples on Mars and returns them to Earth for detailed analysis have been postponed again and again. According to current planning, such a sample-return mission may or may not take place in the second decade of this century. But while they are waiting for this perpetually delayed project to materialize, researchers interested in Martian samples already have at least some material to work with.

A class of meteorites called the SNCs had long been of particular interest to the scientists who study these things. The SNC meteorites were named after Shergotty (India), Nakhla (Egypt), and Chassigny (France), where the first examples of each of the three types were observed falling from the heavens.* The SNCs had attracted attention for two reasons. The first was that, while they differed significantly from one another in terms of their mineralogy, their oxygen isotope ratios clearly indicated that all three types were samples of the same Solar System body. The second was their surprising youth. Isotopic dating placed the age of these rocks at between 1.2 billion years and as little as 160 million years—far, far younger than any other dated meteorites. Most meteorites, remember, come from asteroids, and asteroids are so small that they cooled off and stopped forming new rock shortly after the birth of the Solar System. Therefore, most meteorites produce isotopic ages that are quite close to the 4.56-billion-year age of the Solar System. The SNC meteorites, in contrast, must come from some place that did not cool off (and thus did not become geologically dead) as rapidly as the asteroids. But where might that have been? The isotopic pattern in these meteorites indicated they all came from a common source, but the pattern also indicated that source was neither the Earth nor, by comparison with *Apollo* samples, the Moon. Mars thus seemed the only logical choice, a hypothesis that was confirmed when small bubbles of atmosphere trapped in one of the SNC meteorites were analyzed. Researchers found that the chemical and isotopic contents of these bubbles are in perfect agreement with the composition of the current Martian atmosphere as determined by *Viking* (fig. 10.8). There is now no doubt: the SNCs are samples of Mars that we can hold in our hands and examine at leisure.

How were the SNC meteorites delivered into our backyards? Occasionally, meteorite impacts are violent enough to catapult material into space (as described in chapter 3, our Moon is an extreme example!). Mars being relatively small (its surface gravity is one-third the Earth's)

*The Nakhla fall in 1911 killed a dog, which is the only record of death-by-meteorite.

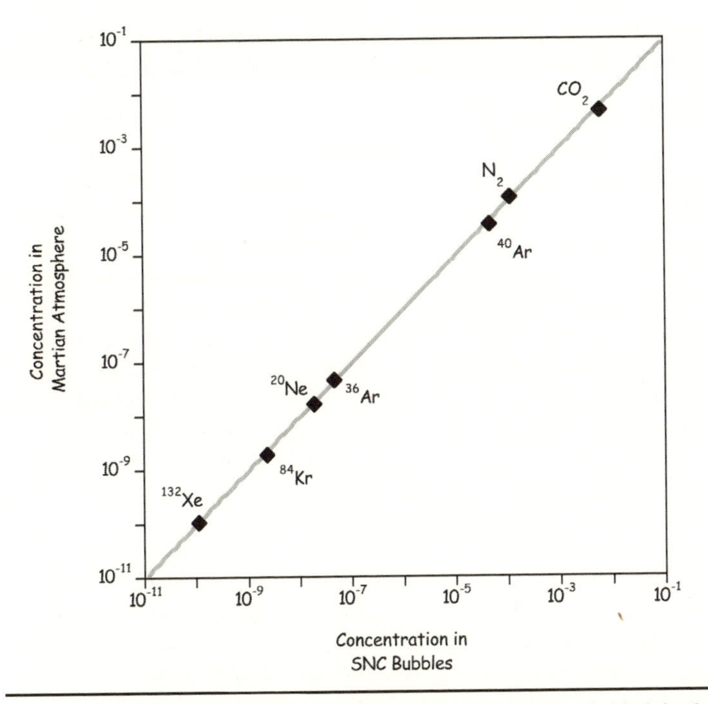

FIG. 10.8. Gas-filled bubbles in one of the SNC meteorites provided the final clue confirming their origin on Mars. The composition of the bubbles is a near perfect match to the composition of the Martian atmosphere as defined by the *Viking* landers.

and having a thin atmosphere, it is not very difficult for an impact to launch Mars rocks into space. And since Mars orbits between the Earth and Jupiter, rocks that escape Mars's gravity and enter solar orbit can be quickly (geologically speaking) perturbed by Jupiter's gravity into orbits that intersect with the Earth's and fall at our feet.

And what does this have to do with our story of life in the cosmos? The link is a Martian meteorite that caused a media frenzy in 1996 and has fueled controversy ever since. In August of that year, NASA scientists claimed in the journal *Science* to have identified evidence (not proof!) of life in the rock named ALH84001 (because it was the first meteorite found in the Allen Hills ice field, Antarctica,* during the 1984 collecting season). ALH84001 is much older than all of the other SNC meteorites:

*Meteorites are easier to identify when they're lying on ice and snow, and thus Antarctica is prime meteorite hunting ground—especially those spots on the continent at which the movement of the ice tends to concentrate any meteorites that have fallen on it.

SIDEBAR 10.1

From Mars to Earth

The advent of conclusive proof that Mars rocks could travel through space to Earth revitalized, in a small way, the panspermia (trans-spermia?) theory that life travels between the stars to seed new planets. If rocks can travel between the planets, could they not bring life with them? Perhaps they can.

Computer simulations of the transit of rocks between Mars and Earth suggest that, on average, a rock blasted from Mars will orbit the Sun several million times before its orbit is perturbed enough that it crosses the Earth's. Studies of the transit time of the SNC meteorites confirm this result. The transit time, or lifetime, of meteorites in space can be determined, because cosmic rays alter the isotopic composition of minerals in measurable ways. When on a planet's surface, a rock is protected from cosmic ray exposure. When it is in space, however, these energetic elementary particles produce characteristic noble gas isotopes in the rock, such as helium-3, neon-21, and argon-38. By monitoring the build-up of these three isotopes in ALH84001, researchers estimated the length of time it spent in space, en route to the Earth, at 16–17 million years.

If rocks can make it from Mars to Earth, could Martian life survive the journey? Living in a crack within a rock would help in one way: it would protect the bug from the deathly effects of solar ultraviolet and, to a lesser but still significant extent, from cosmic rays. Still better, the same computer simulations that predicted a several-million-year mean transit time indicate that rare transits could take as little as a few years. Coupling this with estimates that, on average, several tons of Martian rocks make it to Earth each year, it is possible that Martian life—if there is any—could have hitched a ride and survived the trip. And while it is harder to lift material out against Earth's much stronger gravitational pull, it is probable that Mars and Earth have been engaging in a two-way exchange of material since the origin of the Solar System.

Could Earth life have seeded Mars? Indeed, could Mars have seeded life on Earth? Obviously, short of identifying Earth-like life on Mars we cannot answer this question. But should we ever find organisms on Mars we should keep this point in mind. They may be our long-lost relatives.

---

it solidified almost 4.5 billion years ago and thus represents a piece of Mars's oldest crust. Studies of the radioactive elements produced while it was exposed to cosmic rays in space indicated that it had spent about 16 million years on its trip from Mars to Earth (see sidebar 10.1), and based on the decay of those isotopes (once the rock was on the Earth's surface and spared exposure to any further cosmic rays), ALH84001 is thought to have landed in Antarctica around 13,000 years ago. Like

FIG. 10.9. The ALH8004 meteorite was blasted off the surface of Mars some 16 million years ago, landed in Antarctica about 13,000 years ago, and was collected from the ice in 1984. In 1996, David McKay and colleagues stunned the world by announcing that the 3.6-billion-year-old carbonate globules in the meteorite (upper inset) contained evidence for past life on Mars. This evidence included "nanofossils" (lower inset), submicrometer-sized objects that somewhat resemble Terrestrial bacteria writ very, very small. (Courtesy NASA)

many other meteorites found in or on the Antarctic ice, it is well preserved and has not suffered significant weathering (fig. 10.9).

The evidence used in the original *Science* paper that claimed the presence of microfossils in ALH84001 was essentially the following:

- Organic molecules: Polycyclic aromatic hydrocarbons (PAHs) were detected at fracture surfaces within the meteorite. And while PAHs are relatively common in Terrestrial environments (in diesel smoke, for example), the concentration of PAHs in ALH84001 increased as one went deeper into the rock, suggesting they came with the meteorite and were not contaminants picked up while it sat in the Antarctic ice. PAHs can form via the breakdown of biological organics, and in Terrestrial rocks they are considered a prime signature of biogenic activity.

- Carbonate globules (fig. 10.9): In contrast to the other SNC meteorites, ALH84001 contains spherical inclusions of carbonate (a mineral that is often laid down from aqueous solution) with diameters ranging from 1 to 250 micrometers (thousandths of a millimeter). At an estimated age of 3.5 billion years, these are significantly younger than the surrounding rock and, as judged by several lines of evidence, clearly formed in situ on Mars. Using oxygen isotopic signatures, the *Science* authors estimated that the carbonates were deposited from liquid water at temperatures well below 100°C.
- "Nanofossils" (fig. 10.9): Structures resembling Terrestrial microfossils of cells were found close to these globules. However, they are less than 100 nanometers long, some ten times smaller than the smallest, well-established Terrestrial microbes.
- Unusual iron chemistry: The meteorite also contains small particles of iron sulfide (Fe_2S_3) and magnetite (Fe_3O_4) in close conjunction —minerals of differing oxidation states, which do not normally exist together in equilibrium.

The authors admitted from the start that each of these features might have arisen from nonbiological causes. They concluded, however, that the simultaneous presence of these features in the Martian rock constituted very strong circumstantial evidence for the existence of life on early Mars. Falling in with this conclusion, follow-on studies suggested that the structure of the magnetite crystals is similar to that of the small, intracellular magnetic-sensing "organs" produced by Terrestrial magnetotactic bacteria (which use magnetic beads to sense the Earth's magnetic field and, in turn, to sense the directions "up" and "down"; when you are as small as a bacterium, gravity doesn't provide much of a cue!). No abiological process was then known that could produce these oddly shaped crystals. And others have noted that the sulfur isotope fractionation on the meteorite hints at biological processes as well. Short of a test tube full of living, breathing (well, respiring at least) Martian bacteria, what more could one ask for?

Sadly, though (especially for those of us who'd really like to get our hands on some Martian life to see what makes it tick), the scientific consensus seems to have tilted away from the conclusion that ALH84001 contains signs of life. This tilt was caused by the slow but seemingly inexorable questions raised about each of the original arguments. First, it has been pointed out that carbonaceous chondrite meteorites contain PAHs, and even interstellar clouds of gas. Since it seems unlikely that these PAHs are the products of living processes, it is clear that the PAHs in ALH84001 might also be abiological in origin. This seems all the more likely when the structures of the PAHs are investigated in more detail; PAHs come in a wide variety of molecular structures, and biologically produced PAHs are generally quite diverse. The pattern of PAHs in ALH84001, in contrast, is rather bland and speaks more clearly to abiological processes than it does to life.

Significant controversy has erupted, too, over the original claim that the carbonates were deposited from relatively low-temperature water. On the one side, studies of the meteorite's magnetic properties by Joseph Kirschvink and of the oxygen isotope ratios of the carbonates by Edward Stolper, both of the California Institute of Technology, have been said to support the low-temperature claim. On the other side, scientists Edward Scott of the University of Hawaii and Harry McSween of the University of Tennessee, Knoxville, among others, have argued equally forcefully for deposition at temperatures far too high to support even extremophilic life. Similar controversy has focused on the "nanofossils." While there are some (widely disputed) claims of nanometer-sized bacteria here on Earth, the possibility of such organisms is generally dismissed. The reason is that the volume of these putative cells would be less than one-thousandth the volume of a typical bacterial cell, and thus certainly too small to contain the metabolic machinery of even the simplest free-living organisms on Earth. Steve Benner has pointed out, though, that ribosomes take up much of the space inside a bacterium, and if the Martian organisms date back to the RNA world (before the invention of ribosomes and, with them, proteins), it is just conceivable that they could be as small as the putative fossils. Still, while each of these points may be debatable—and are being vigorously debated even now, a decade after the original announcement—the first three lines of evidence initially put forth as evidence of life in ALH84001 have generated little in the way of scientific consensus.

In contrast, the observation of odd bits of the mineral magnetite in ALH84001 was for some time rather better received. For example, the truncated hexa-octahedral shape of these crystals was considered probably the best evidence for fossil life. In terms of size and composition, the fossil magnetite looks exactly like the magnetite beads from magnetotactic bacteria, so much so that, if they had been found on Earth, they would be considered uncontroversial "magnetofossils." In fact, Kirschvink, the prime proponent of this argument, has claimed that these fossils make the case for life on Mars some 3.7 billion years ago much more well-founded than the case for life on Earth at the same time! Unfortunately, though, a team led by M. S. Bell of the University of Houston, Texas, has shown that identical magnetite particles can be synthesized via the thermal decomposition of iron carbonates, a decidedly abiological process. Thus, while the question of whether ALH84001 contains authentic traces of Martian life ultimately remains unresolved, the broad scientific consensus is shifting toward "dead as a doornail."

Astrobiology in the Outer Solar System

While much of the public attention on Solar System exploration has focused on Mars, the renewed optimism in astrobiology has also raised the profile of a few other solid bodies in our neighborhood, especially the icy moons of the gas giants. Jupiter's satellite Europa, with its ice shield covering large amounts of a conducting liquid presumed to be saltwater, is now a prime candidate for extraterrestrial life, as we discussed in the previous chapter. Current estimates, however, suggest that Europa's ocean lies beneath a 10- to 100-kilometer thick crust of ice; it is, unfortunately, going to be quite some time before anyone lands a craft on this icy moon that can melt its way down to have a look. Meanwhile, however, NASA's *Cassini* spacecraft arrived in orbit around Saturn on July 1, 2004, together with its lander *Huygens,* destined for Titan. And while this enormous moon is considered unlikely, but not impossible, as an abode for life, its study by *Cassini* and *Huygens* is advancing our understanding of prebiotic chemistry.

Titan is the only satellite in the Solar System with a thick atmosphere (fig. 10.10). In fact, the atmospheric pressure at the surface of Titan is 1.5 times that at Earth's surface, which, given Titan's much weaker gravity, corresponds to an atmospheric density some three times ours. The Titanian atmosphere consists predominantly of nitrogen gas, with only a few percent methane, the simplest organic molecule. Under the influence of ultraviolet light from the Sun, the methane and nitrogen in Titan's atmosphere react to form a thick haze of higher-molecular-weight organics. The haze is so thick, in fact, that the first flyby spacecraft, *Pioneer 11* and the two *Voyagers,* entirely failed to see the moon's surface, and *Cassini* is equipped with a radar imaging system to peer through the smog. The photolysis reaction that produces this haze is so rapid that the resulting loss of hydrogen to space (à la Venus's loss of water; see chapter 3) would deplete Titan's atmospheric methane in only 10 million years, unless something was replenishing it. Initially it was thought that, just as our oceans keep our atmosphere humid, a Titanian ocean of methane might explain the presence of this gas in the atmosphere. More recently these oceans were downgraded to, at best, lakes. Still, it seems that Titan has just the right range of temperatures for methane to exist as a liquid, solid, or gas, and thus methane may drive a weather cycle on Titan analogous to our water-driven weather cycles. Consistent with

FIG. 10.10. The haze-filled atmosphere of Titan, Saturn's largest moon, is thought to be a 4-billion-year-old laboratory of prebiotic chemistry. (Courtesy NASA/JPL)

this, the first *Cassini* radar images of Titan's surface suggest that the moon's surface bears very few impact scars, perhaps because craters erode rapidly under the onslaught of methane rains and rivers. More startling still, the *Huygens* probe snapped pictures of what look like river channels and lakebeds (fig. 10.11) during its descent to land on the cobble-covered surface of the moon in early 2005 (fig. 10.12).

As fascinating as the surface geology of Titan is turning out to be, it is Titan's atmospheric chemistry that is of interest to the astrobiology community. The reason is that Titan's atmosphere, while very cold, seems to be a rough analog of the early Earth as postulated by

FIG. 10.11. During its two-and-a-half-hour descent to the surface of Titan, the *Huygens* probe snapped these photos of what may be river channels carved by methane rains. Titan seems to have a complex, liquid-methane-based fluid cycle reminiscent of the Earth's hydrological cycle. Does it host life as well? Probably not. But it no doubt hosts fascinating chemistry—chemistry that could tell us a great deal about our own origins. (Courtesy NASA/ESA)

Harold Urey (save its being so cold that it lacks water and other oxygen-containing compounds). That is, the atmosphere is reducing and nitrogen-rich. Under the influence of ultraviolet radiation from the Sun (and—who knows?—perhaps lightning in those methane clouds), these dominant atmospheric components, so it is thought, must be reacting to produce life's precursors, which form the thick haze and no doubt rain down on the surface. Over billions of years, perhaps the type of organics that Stanley Miller observed fifty years ago have accumulated hundreds of meters thick on the surface of this frozen world. The *Huygens* probe, alas, was not equipped to search for such molecules on the surface (although it did survive for more than an hour after impact and detected such large volumes of methane emanating from the surface below it that the ground must be saturated with the stuff, much as sand on the beach can be saturated with water). Still, some of the *Cassini* flybys

FIG. 10.12. After a seven-year, 5-billion-kilometer trip, the *Huygens* probe landed on Titan's frozen ground to become, for an hour and a half, humanity's most distant outpost. In the distance, beyond scattered cobbles of water frozen to rock hardness by the deep cold, lies the Titanian horizon. (Courtesy NASA/ESA)

have been close enough that the orbiter's mass spectrometer has detected high-molecular-weight hydrocarbons and nitriles (hydrocarbons containing a carbon-nitrogen triple bond) in the moon's outermost atmosphere. As scientists continue to chug through the *Huygens* data, and as *Cassini* continues to fly by Titan (several dozen passes are planned), we will no doubt be hearing much more about the chemistry of this fascinating moon.

The Search for Life beyond the Solar System

As we described in the previous chapter, recent years have seen the identification of more than a hundred extrasolar planets, with the rate of discovery rapidly increasing. All of the currently identified extrasolar planets, however, are gas giants, and most are exceedingly close to their companion stars and thus exceedingly unlikely places for life to have gained a foothold. And yet these discoveries have given researchers some confidence to renew their search for terrestrial planets orbiting in habitable zones.

While such planets would be too small to produce significant Doppler shifts of the sort that yielded information about the first hundred extrasolar planets, even an Earth-sized planet can be observed directly if it happens to pass in front of its star relative to our vantage point. Such "transit events" can easily be detected by the slight dip, amounting to only a fraction of a percent, in the star's brightness, provided astronomers are equipped with a good telescope and lots of patience (the chance of any one planet crossing directly between us and its star is quite small). For example, in April 2003, researchers at the European Southern Observatory in Chile discovered a planet that is less massive than Jupiter and obscures its star once every 28.5 hours (such a rapid orbit implies it is close to its companion star, so close that the planet must glow red hot). The fact that such observations are possible with ground-based telescopes suggests that space-based telescopes and interferometers fully dedicated to the planet search should be able to detect even Earth-like planets in Earth-like orbits. Several such instruments are now in the works.

The first device capable of spotting rocky planets as small as a few times the size of Earth will be the *Corot* probe, a collaborative effort of the French and the European Space Agency (ESA) that is due to be launched into Earth orbit in early 2006. *Corot* will be equipped with a 30-centimeter telescope designed to survey fluctuations in the brightness of some 60,000 stars, able to detect not only stellar weather, such as sunspots (which also cause small—few percent—changes in a star's brightness), but also (scientists hope) planetary transits. The first-generation instrument should be able to detect the dip caused by Jupiter-sized planets. Follow-on plans by the ESA include missions to expand *Corot*'s survey with larger space-based telescopes that will eventually provide the resolution necessary to detect planets similar in size to Earth or Venus that are orbiting nearby stars.

An even more ambitious project to survey our galactic neighborhood is ESA's *Gaia* probe, which is to be launched around 2010. To be located at the outer Lagrange point, the point at which the Sun's and the Earth's gravity are equal, the probe will orbit the Sun in step with its home planet, but 1.5 million kilometers farther out. Using a combination of three telescopes (two for positioning a star and one for analyzing its light), it will survey up to a billion stars, and might detect thousands of extrasolar planets. Thus by 2015 we should know conclusively whether Earth-sized planets are common in our galaxy. But will we know anything about whether they are inhabited?

Considering how difficult it is for us to prove or disprove the existence of life on Mars, it will clearly be a challenge to do the same for extrasolar Earth-like planets, if any are found. However, the transit method allows researchers to investigate the spectroscopic properties of the planet's atmosphere while some of the starlight we receive passes through it. (This has been performed on one of the hot, closely orbiting Jupiter-sized planets we discussed in the previous chapter, which was found to contain an atmosphere rich in sodium gas. It's damn hot there!) Even more promising, with the right equipment it may be possible to directly measure the spectra of extrasolar planets (no easy task when you consider that they shine by reflected starlight and thus are a billion times dimmer than the star they orbit). Such observations will be included in the mission of NASA's space-based twin observatory *Terrestrial Planet Finder* (TPF), which, with a good dose of political and technical luck, will be launched within the next ten to fifteen years. It comprises two separate parts. *TPF-C,* due to launch around 2014, will carry a telescope operating at optical wavelengths and equipped with a coronagraph—that is, a small disk that blacks out the glare of the star that the telescope is pointed at—to facilitate the observation of less radiant objects nearby, such as planets. The second part of the mission, *TPF-I,* will be a large baseline infrared interferometer, to be launched in cooperation with ESA before 2020. At around the same time, ESA plans to launch another mission designed to search for habitable planets, under the name *Darwin.* It will consist of six infrared telescopes in an arrangement 100 meters across and positioned at the outer Lagrange point. With forty times the resolving power of the Hubble telescope, it is hoped that *Darwin* will be able to analyze light from terrestrial planets as distant as 50 light-years. Thus, if there are living planets in our galactic neighborhood, it seems reasonable to hope that we may find them within the next couple of decades and even measure their spectra. And will these missions be able to detect unambiguous signatures of life? Perhaps not. But with long integration times, so it is believed, they would be able to detect, for example, ozone. And ozone is a hallmark of a dense oxygen atmosphere, which, with the caveats we've described, can be considered a strong indication of life.

SETI: The Search for Extraterrestrial Intelligent Life

The search for extraterrestrial life, as we have described it thus far, consists of the two main approaches of finding habitable celestial bodies and checking for any chemical traces of life, past or present, primitive or evolved. It left out one significant aspect of the search, namely the search for intelligent life—that is, for any kind of life form advanced enough to be able to communicate with us *directly* across space. Myths and a lot of bad television aside, no such life has been detected by the time of writing, so the contents of this last section of the book are bound to be based on probabilities, speculation, and philosophy. But the question still has to be asked: is anybody out there and, if so, why haven't we heard from them?

A fundamental problem here is that, while some of our technology may seem pretty amazing (especially to those of us who won't see thirty again), human technology is extremely limited in comparison to the dimensions of the Universe. And we're talking about serious limitations, not obstacles that we can overcome next year or even in ten years time. For example, it has been speculated that the first person to set foot on Mars has already been born. In contrast, the first person to travel to α-Centauri (at 4.3 light-years away, our nearest stellar neighbor) and come back alive has certainly not been born yet, and the people who will boldly go to visit other galaxies will not be born for a long time, if ever. Given the enormous times and distances that would be involved in interstellar travel, any hands-on exploration we conduct will be limited to the Solar System for the foreseeable future.

The limitations imposed by the vast distances of space are somewhat relieved when we consider the search for civilizations advanced enough to communicate by light or radio waves. As the tireless astrobiologist Carl Sagan illustrated in his novel *Contact,* the first major TV transmission—the opening of the Olympic Games in Berlin in 1936—is still winging its way into space at the speed of light (implying that it has now been broadcast to everybody living within a radius of 70 light-years). Of course, that broadcast was pretty weak and would be well-nigh impossible to detect at truly galactic distances. In contrast, it is said that the giant dish antenna at Arecibo, Puerto Rico, the Earth's most powerful radio telescope (which is also fitted with a transmitter for radar studies), could hear transmissions from a similar antenna from clear across our galaxy (albeit at a paltry one bit per hour bandwidth). One can only speculate what civilizations with much more advanced technologies might do if they set their minds to it. Quite probably, they could communicate even from beyond our own galaxy. Therefore, in the "search for extraterrestrial intelligence," widely known as SETI, the entire Universe is our haystack.

First attempts to search for radio signals from intelligent extraterrestrials date back to 1960 with Project Ozma (named after a princess from Frank Baum's imaginary land of Oz). Initiated by the astronomer

Frank Drake, then at Cornell University in upstate New York, Project Ozma used a 24-meter radio telescope at the National Astronomy Observatory in Green Bank, West Virginia, to listen for messages from the stars τ-Ceti and ϵ-Eridani, at a frequency of about 1,420 GHz (gigahertz—billion hertz). This frequency is in the so-called watering hole, which is between the frequencies at which hydrogen atoms and hydroxyl radicals emit radio waves, and thus, it would seem, a logical place for at least water-based life to broadcast messages, if it intended them to be found amid the myriad of frequencies on the radio dial. Drake recorded for 150 hours over the course of several months and, at one point, found some nonrandom signals that seemed intelligent in nature. This must have surprised even him, though, for despite being perhaps an extreme optimist with regard to the number of communicating civilizations in our galaxy (see sidebar 10.2), even Drake estimated that the chance of any single, seemingly suitable star harboring intelligent life is much less than 1 in 10,000. His surprise, though, couldn't have lasted long, as the signals turned out to be emanating from a then secret military radar. No apparent intelligent extraterrestrial signals were recorded.

By the 1970s SETI proponents had grown more active. In 1971 John Billingham of NASA's Ames Research Center in California authored a detailed study of the feasibility of building an "antenna farm," called Cyclops, comprising a thousand 100-meter dishes. Cyclops, Billingham argued, could detect routine television and radio signals, not to mention intentional attempts to communicate with us, from any of a large number of neighboring stars. The Cyclops proposal didn't get very far, perhaps because of its $10 billion price tag. Four years later NASA judged that the relevant science and technology had become mature enough to give it another go, and the next year SETI research programs were initiated at Ames and at California's Jet Propulsion Laboratory. But it was not to last; serious congressional unease about the expense of "a silly search for aliens that was unlikely to yield results" hampered efforts to fund SETI, until Sagan stepped into the fray. A well-known and well-respected astronomer and popularizer of science, Sagan was able to convince key Senate players of the merits (and relatively low cost) of SETI, such that Congress reinstated NASA's SETI funding in 1983, and NASA began building hardware in 1988. The SETI program, rather seriously named the High Resolution Microwave Survey in an attempt to get past the "giggle factor," finally began searching the skies in 1992. Sadly, it was not to last. The congressional rhetoric over "little green men" heated up significantly in the early 1990s, and the program was canceled in 1994. "This hopefully will be the end of Martian hunting season at the taxpayer's expense," commented Richard Bryn (quoted by Garber, 1999), a

SIDEBAR 10.2

Weighing the Probabilities

Will the SETI attempts ever find intelligent life out there? Assessments that focus solely on the number of stars in our galaxy generally lead to the conclusion that there must be somebody somewhere. On the other hand, Enrico Fermi (1901–54), the Italian-American physicist and 1938 Nobel laureate in physics, argued that we seem to be alone.

Fermi calculated that, even if their top speed was limited to but a small fraction of the speed of light, civilizations with even a modest amount of rocket technology could colonize the entire Milky Way within, say, a few tens of million years. His argument was that, even if it took each new colony planet half a million years to set up two colonies of its own, this exponential growth would lead to more colonies than there are stars in just 20 million years. And while that may seem like a long time, it is extremely short compared with the more than 10-billion-year age of our galaxy. Clearly, Fermi realized, aliens had had plenty of time to colonize the whole galaxy. So, then, if there are a lot of aliens out there (as almost everyone assumes), why aren't they here?

Fermi argued that the fact that aliens don't seem to be hanging out with us here on Earth (the tabloid stories aside) strongly contradicts the assumed existence of intelligent life elsewhere in our galaxy, a problem that came to be known as the Fermi Paradox. And it's a hard paradox to break; you can argue that the aliens can move at 10% of the speed of light or at 1%, and you still come up with more or less the same answer. Namely, that the entire galaxy should be colonized over a period vastly shorter than its age. That it has not been implies that we are likely alone.

On the other side of the argument are many astronomers, who are generally much more optimistic about these things. A prominent member of this camp is Frank Drake, who in 1961 noted that the number of extraterrestrial civilizations we can communicate with will be given as:

$$N = R^{*} f_p \, n_e f_l f_i \, f_c L$$

where:

R^{*} = rate with which suitable (Sun-like?) stars are formed in our galaxy (per year)
f_p = fraction of these stars that have planets
n_e = average number of planets per star with planets that could support life
f_l = fraction of those "qualifying planets" that actually develop life
f_i = fraction of planets with life that develop intelligent life
f_c = fraction of intelligent life forms that are willing and able to communicate
L = average lifetime of a communicating civilization (in years)

Most of these variables, though, are very much unknowns. We can probably estimate the first parameter, R^{*}, to within a factor of two or so, but after that we begin to lose almost all contact with known reality. Nevertheless, many people, starting with Drake, have estimated N and used the Drake equation as a sounding board with which to describe their optimism or pessimism about life elsewhere in our Universe.

Drake, himself, remains an optimist (you'd have to be to persevere in SETI research for more than forty years!). His estimates of the various parameters place the value of *N* at about 10,000 communicating civilizations in the Milky Way alone, which makes SETI seem like a reasonable effort, but also essentially reinforces the Fermi Paradox. But is Drake's optimism well founded? As we've discussed in detail in this book (most notably in chapters 4 and 5), we do not yet understand how life arose on Earth in anywhere near enough detail to support such an optimistic scenario. Admittedly, since we do not know how life arose here we cannot rule out the possibility that f_l is near 1, but the very best theories we currently have concerning the origins of life suggest that f_l might be *tens or hundreds or thousands of orders of magnitude less than 1.* Similarly, since the anthropic principle requires that intelligent life form before we can discuss the probability of intelligent life forming, all we know about f_i is that it is not zero. Its value too, however, may be infinitesimally close to zero and, at the very least, seems likely to be far lower than the near unity assigned to it by Drake.

Lastly, as Carl Sagan pointed out, advanced civilizations may tend to eradicate themselves more rapidly than they can send out colonies via interstellar travel: considering the ratio of rockets built for killing Earthlings to the number built for sending them to other planets, it may well be that advanced civilizations are more likely to self-destruct than to travel to the stars. Perhaps less pessimistically, even here on Earth we have started to use up radio-frequency bandwidths (for cell phones and the like) so rapidly that, within a few years, the radio signals emanating from our planet may look much more like white noise than like a sign of intelligent life. If so, *L* could also be infinitesimally close to zero (the fifty-year or so span during which human radio sources have been beaming intelligible "we are here" signals across the heavens is negligibly small compared with cosmological ages). The range of values that each of Drake's parameters could adopt is so great that, despite the huge number of stars in the Universe, current scientific knowledge is *entirely consistent with* N = *1*. That is, Fermi was right and we are alone.

prominent congressional SETI critic. After less than two years of active searching, NASA's SETI program was dead.

But all is not lost. When the NASA project ended, the SETI Institute was founded in Mountain View, California, under the direction of none other than Frank Drake, the astronomer who had started the SETI ball rolling back in 1960. To this day, the SETI Institute continues to play a major role in SETI research and education, entirely with private support. Recently it has been involved in the design of the Allen Telescope Array, which was largely funded by Paul Allen (cofounder of Microsoft) and is being built and operated in collaboration with the University of California, Berkeley. A main feature of this new array is that it can be scaled up gradually. Currently coming in at just under three dozen 6.1-meter radio telescopes, the final version of 350 dishes is expected to be complete in 2008. The Allen Telescope Array is designed for dual use, not only to search for radio signals from advanced civilizations across our galaxy, but also for a wide range of basic radio astronomy studies.

The SETI endeavor is, however, not just about having a sufficient number of antennae. Sifting through the huge amounts of radio data, looking to extract intelligent signals from the noise, is also a major challenge. A pioneering solution to that problem was provided by the SETI@home project, which farmed out the automatic data-sifting process to thousands of personal computers as screen savers. The idea, which since then has also been used in the life sciences, is that PCs that are always on and always online (e.g., most PCs in universities) provide an enormous computational resource that can be exploited, during the computers' idle times, for SETI data processing. To date, the computers connected to the SETI@home project have accumulated 2.3 *million years* of CPU time, and the project continues to accrue computer time at a rate of about 1,000 CPU-years *per day*! So far, though, despite this massive firepower, all of the SETI searches have recorded only noise. The occasional "unexplained signal" has always turned out to have a Terrestrial or astronomical explanation. But the search continues.

Conclusions

But what about the planet we visited at the beginning of this chapter—the one that showed the intriguing indications of chemical disequilibrium that seemed to point to life? The scientists in control of this spacecraft concluded that the planet was indeed inhabited—even, probably, with intelligence (the regular radio-frequency pulses seemed too regular to be of natural origins). The scientists, led by Carl Sagan, excitedly wrote up this work and published in the prestigious British journal *Nature,* which ran the story on its cover on October 21, 1993, under the caption, "Is there life on Earth?"

So, ahem, yes, this question may look silly at first glance, as we wouldn't be here to discuss it if the answer was no. The spacecraft in question, NASA's *Galileo,* had to swing by its home planet for a gravity assist on its way to Jupiter, with all its instruments blazing away (for calibration purposes). Sagan and his coworkers realized that the flyby provided an ideal opportunity to find out how difficult it would be to detect life on a planet that had been teaming with it for billions of years. The resulting paper is interesting in that it reveals the difficulties in obtaining unequivocal evidence for life on a planet during a flyby. In the end, however, the evidence of serious chemical disequilibrium in combination with the radiation and absorbance characteristics of the planet was counted as sufficient to conclude that there is life on Earth. Thus the result obtained by the spacecraft swinging by our planet confirmed the "ground truth" that we knew all along, but still managed to teach us a few lessons about how difficult it is to detect life in the Universe.

Further Reading

The search for life on Earth. Sagan, C., Thompson, W. R., Carlson, R., Gurnett, D., and Hord, C. "A search for life on Earth from the Galileo spacecraft." *Nature* 365 (1993): 715–21.

Life on Mars (general). Chambers, Paul. *Life on Mars.* New York: Stirling Publishing, 1999.

Evidence for life in ALH84001. McKay, D. S., Gibson, E. K., Jr., Thomas-Keprta, K. L., Vali, H., Romanek, C. S., Clemett, S. J., Chillier, X. D., Maechling, C. R., and Zare, R. N. "Search for past life on Mars: possible relic biogenic activity in Martian meteorite ALH84001." *Science* 273 (1996): 924–30.

SETI. Shklovskii, I. S., and Sagan, C. *Intelligent Life in the Universe.* New York: Dell, 1966; Garber, S. J. "Searching for good science: the cancellation of NASA's SETI program." *Journal of the British Interplanetary Society* 52 (1999): 3–12; also, visit www.seti.org.

INDEX